Pañcāstikāyasāra of Kundakunda

By the Same Author

Advaita Vedanta and Jaina Advaita (2024)

Ratnakaranda Shravakachara:
Spirtual Life of the Householder (2022)

Perspectives in Jainism (2017)

The Essence of Jaina Scriptures:
Pravachanasara of Kundakunda (2014)

The Art and Science of Self-Realization:
Purushartha-Siddhyupaya of Amritachandra Suri (2006)

Fundamentals of Jainism (2005)

Salvation through Self-Discipline:
Niyamsara of Kundakunda (2005)

The Enlightened Vision of the Self:
Svarupa Sambodhana of Akalanka Deva (ed.) (2005)

Spiritual Enlightenment:
Paramatma Prakash of Yogindu Deva (ed.) (2000)

Spiritual Insights: Ishtopadesh and
Samadhi Shatak of Pujyapada (ed.) (2000)

Pañcāstikāyasāra of Kundakunda

Building of the Cosmos

Edited, Translated and Commentary by
Jagdish Prasad Jain

PRINTWORLD
Publishers of Indian Traditions

Cataloging in Publication Data — DK

[Courtesy: D.K. Agencies (P) Ltd. <docinfo@dkagencies.com>]

Kundakunda, author.

Pañcāstikāyasāra of Kundakunda : building of the cosmos / edited, translated and commentary by Jagdish Prasad Jain.

pages cm

English and Sanskrit (Sanskrit in Devanagari and roman).

Includes bibliographical references.

ISBN 9788124612446

1. Jaina philosophy – Early works to 1800. 2. Knowledge, Theory of (Jainism) – Early works to 1800. 3. Kundakunda. Pañcāstikāya. I. Sadhak, Jagdish Prasad Jain, editor, translator. II. Kundakunda. Pañcāstikāya. III. Kundakunda. Pañcāstikāya. English. IV. Title.

LCC B162.5.K86 2024 | DDC 181.044 23

ISBN: 978-81-246-1244-6

First published in 2024

Printed and published by:

D.K. Printworld (P) Ltd.

Regd. Office : "Vedaśrī", F-395, Sudarshan Park

ESI Hospital Metro Station, New Delhi - 110015

Phones : (011) 4172 2626, 98713 94815

e-mail : indology@dkprintworld.com

Web : www.dkprintworld.com

Contents

Abbreviations

BA	*Bārasa Anuvekhā*
JS	Jayasena
NS	*Niyamasāra*
PKS	*Pañcāstikāyasāra*
PS	*Pravacanasāra*
RKS	*Ratnakaraṇḍa Śrāvakācāra*
SSK	*Samayasara Kalash*
SS	*Samayasāra*
TS	*Tattvārthasūtra*

Introduction

Pañcāstikāyasāra (*PKS*) is one of the important works of Kundakunda, who was an outstanding philosopher, a versatile genius, a celebrated literary figure, an original thinker, a doyen of saints and a great religious teacher. He lived in the first century BCE.

Kundakunda's name is revered next only to Lord Mahāvīra and his apostle Gautam Gandhara. *PKS* deals with Jaina metaphysics or ontology and ethics, leading one to the path of salvation.

A number of commentaries on it in Sanskrit by Amṛtacandra, Brahmadeva, Devjita, Jayasena, Jñānacanda, Mallisena and Prabhācandra are known. Those of Amṛtacandra and Jayasena are already printed and are well known. There is a commentary by Bālacandra in Old Kannada; but it has not been published. Some expositions of this work in India have been composed by Hemarāja (Saṁvat 1700), Rājamalla (Saṁvat 1716), Hīrācandra (Saṁvat 1718) and Vidhicanda (Saṁvat 1891) and more recently by Lalaji Nyayatirtha (1989-90 CE/Saṁvat 2046-47).

Like other works of Kundakunda, the *Samayasāra* (*SS*) and *Pravacanasāra* (*PS*), *PKS* too contains the Sanskrit commentaries of both Amṛtacandra and Jayasena. As with other works, Jayasena's recension of the *Pañcāstikāyasāra* also contains additional verses than those in Amṛtacandra's recension of the *Pañcāstikāyasāra*.

As the name itself suggests, the *Pañcāstikāyasāra* deals with the five (*pañca*) cosmic constituents (*astikāyas*). The term *astikāya* is a compound name made up of *asti* and *kāya*, which respectively means existing or existent and of extensive magnitude. *Astikāya* therefore means a real thing that has extensive magnitude, i.e.

capable of occupying space. The term *kāya* implies relation to many space-points.

The five *astikāyas* are eternal, uncreated and of extensive magnitude (*PKS* 4). Moreover, whatever things have the essential nature of manifesting themselves severally through their numerous qualities and modes are the *astikāyas*. These fill the three worlds – being the constituent parts of the world (*PKS* 4).

The *astikāyas* are five in number, viz. *jīva* (soul), *pudgala* (matter), *dharma* (principle of motion), *adharma* (principle of rest) and *ākāśa* (space). These five *astikāyas* build up the cosmos. Space and matter are distinctly extended reals. *Dharma* and *adharma* are indirectly related to space. Their operation is in space and is limited by *lokākāśa*. Thus, they may also be considered as related to space. Lastly, life is generally associated with body, the organic body is constituted by *pudgala*. *Jīva* is operating in a condition by such a physical medium. In a way, therefore, *jīva* is also related to space. These five existences, which have spatially either directly or indirectly, are the five *astikāyas*. These are the consummate elements of the universe or the world. Of these five *astikāyas*, *dharma* and *adharma* are devoid of any kind of direct causal potency. Their function is purely external and indirect.

Kāla (time), though not an element of the physical universe, may be mentioned here. Since change and motion are admitted to be real, time also must be considered real. The reals are absolute time as contrasted with the relative time, that is constituted by simple elements known as *kālāṇus* (instants). Instants, points and atoms are the characteristic conceptions of Jaina thought and in this respect it has a wonderful collaboration from the field of modern mathematics.

The Jaina thinkers in distinguishing time from the five *astikāyas* made use of an important idea. *Astikāya* is spatially of extensive magnitude. This extensive magnitude is denoted by a technical name, *tiryak-pracaya* (horizontal extension).

When the simple elements, say, the points are so arranged in a series where each term is an item also in another series, we must have the two-dimensional series which will correspond to the surface or extension. Wherever there is such a *tiryak-pracaya*, we have *astikāya*. But *kāla* has only *ūrdhva-pracaya*. The elements are in a forward direction. The series is mono-dimensional or of a linear order. Therefore, *kāla* has no extension either directly or indirectly. Hence, it is not an *astikāya*. Though it is not an *astikāya*, it is distinctly a real entity which accounts for changes in other things.

The five *astikāyas* and time together constitute the six *dravyas* (substances). The term "substance" is suggestive of something that subsists or persists in spite of the changes. The term *dravya*, thus, denotes any existence which has the important characteristic of persistence through change.

The Jaina conception of reality excludes both permanent and unchanging real of the Parmenides type (or Vedānta) and also the mere eternal flux of Haraclitus (or Buddhists). And unchanging permanent and mere change without substratum are unreal or impossible abstractions. The Jaina system admits only the dynamic reality (*dravya*). The very term *dravya* signifies *dravyatva* (that which by nature flows towards its modes) (*PKS* 9).

Substance (*dravya*) is defined by Kundakunda as follows:

> That which, whilst it does not forsake its innate nature, is characterized by origination (*utpāda*), decay or annihiliation (*vyaya*) and permanency or stability (*dhrauvya*), which possesses qualities (*guṇa*) and modes or modifications (*siddha*).
>
> – *PS* 95; *PKS* 10; *TS* V.30 and V.38

Thus, *dravya* forms the triple nature of the real. To emphasize the underlying identity alone would end the Vedāntic conception of this real as *Brahman*. To emphasize the change alone would result in the Kṣaṇikavāda of the Buddhists, the reality as a stream of discrete and momentary elements. The (Jaina) concept of *dravya* reconciles both these aspects and combines them into an organic unity. It is an identity expressing through difference, a permanency continuing through change. It corresponds to the modern conception of organic development rather in its Hegelian aspect.

According to Kalipada Mitra,

> the stages of origination and annihilation are like the thesis and anti-thesis of Hegel having a tendency toward stability, which means nothing other than synthesis at a particular stage of the continuous developmental process ready to make room for a fresh origination or a new stage.[1]

After describing substances in general in twenty-one verses (*PKS* 6-26), Kundakunda examines in detail *jīva, dravya* (soul substance) in forty-seven verses (*PKS* 27-73), *pudgala dravya* (matter substance) in nine verses (*PKS* 74-82), *dharma–adharma dravya* in seven verses (*PKS* 83-89) and *ākāśa* (space) in fifteen verses (*PKS* 90-104) in four chapters of Book 1.

In *Pañcāstikāyasāra* 105, again the obeisance to Lord Mahāvīra, the cause of liberation, Kundakunda describes nine basic principles of life (*tattvas*) or categories (*padārthas*) [the first two primary categories of *jīva* (soul, i.e. conscious being) and *ajīva* (non-conscious, non-living thing)], including matter (*pudgala*) and the seven derivative native categories of merit or virtue (*puṇya*) sin or vice (*pāpa*), influx (*āśrava*), stoppage (*saṁvara*), shedding (*nirjara*) and bondage (*bandha*) of *karmas* and liberation (*mokṣa*) as also the path of liberation (in two verses – *PKS* 106-107). The nine categories are mentioned in *Pañcāstikāyasāra* 108 and discussed in eight chapters (in *PKS* 109-175).

---

[1] Kalipada Mitra, 1939, "The Jaina Theory of Existence and Reality", *Indian Culture*, p. 323, January.

Pañcāstikāyasāra

Text, Transliteration, Translation and Commentary

Benedictory Poetic Composition (Maṅgalācaraṇa)

1

इंदसदवंदियाणं तिहुअणहिदमधुरविसदवक्काणं।
अंतातीदगुणाणं णमो जिणाणं जिदभवाणं॥

iṅdasadavaṅdiyāṇaṁ tihuaṇahidamadhuravisada-
vakkāṇaṁ ।
aṅtātīdaguṇāṇaṁ ṇamo jiṇāṇaṁ jidabhavāṇaṁ ॥

I pay obeisance to the Omniscient Lords (Jinas), possessed of infinite attributes; the conquerors of transmigration (*bhava, saṁsāra*), i.e. who are beyond the influences of transmigration; worshipped by hundred Indras; and whose utterings, speech or words are sweet and clear or evident and beneficial to the three worlds.

Commentary by Amr̥tacandra

The three worlds are upper, middle and lower. The words of Jinas are described as beneficial (*hitakara*) because they enable the people to attain the unrestricted or uninterrupted (*nirbādha*) purification of the *ātma-tattva*, i.e. the spiritual principle (*tattva*) of the Self or pure and perfect Self.

2

समणमुहुग्गदमट्ठं चदुग्गदिणिवारणं सणिव्वाणं।
एसो पणमिय सिरसा समयमियं सुणह वोच्छामि॥

samaṇamuhuggadamaṭṭhaṁ caduggadiṇivāraṇaṁ
saṇivvāṇaṁ ı
eso paṇamiya sirasā samayamiyaṁ suṇaha vocchāmi ॥

I bow my head with reverence to the scripture (*samaya*, i.e. *āgama*) (*PKS* 1 AC) that is revealed by the Omniscient Lord (Jina), which is the cause of averting or the removal of the four conditions of existence (*caturgati nivāraṇa*) and leading to liberation (*nirvāṇa*): Listen! I describe the same truth in this work.

Commentary

The four *gati*s (conditions of existence) are: (1) hell (*naraka*); (2) plant and animal life (*tiryañc*); (3) human being (*manuṣya*); and (4) the celestial or the heavenly beings (*deva*s). These four *gati*s constitute *saṁsāra*.

The *āgama* (scripture) is saluted for two reasons: its origin and fruit. It is the word of God revealed by the Omniscient Lord (*sarvajña*). The fruit is that it saves the soul from the four *gati*s of *saṁsāra*, the transmigratory existences of births and deaths, and leads the Self to emancipation (*nirvāṇa*), the unconditioned state of the pure and perfect Self.

Here, *samaya* is used in the sense of the scriptures (*āgamas*). In his introductory remarks to *Pañcāstikāyasāra* 2, Amr̥tacandra categorically states that "*samaya* alone is *āgama*".

In *Pañcāstikāyasāra* 2, Kundakunda promises to describe the scripture. He states:

> I bow to the scripture (1) that which describes the substances (*artha* or *padārtha*) uttered from the mouth of or revealed by the Omniscient Lord (called *saman* (*śareman*)) [hence it is authentic and fruitful (*Pañcāstikāyasāra* 2 AC)]; (2) that wards off or averts transmigration into four conditions of existences (*caturgati*, viz. human, celestial, hellish and animal); (3) that leads to *nirvāṇa* (liberation). Listen! I describe such a *samaya* (scripture).

Commentary by Amṛtacandra

In his commentary on this verse, Amṛtacandra observes:

> Due to the above reasons, the scripture is said to be fruitful or productive of results (*saphala*); the realization of such a pure Self principle (*śuddha ātma-tattva*) is fruitful due to being the customary cause of *nirvāṇa* (liberation) characterized by the attainment of freedom or liberation.

That by which one properly knows the meaning and sense of the word is *samaya*.[1]

Samaya is used in the sense of *āgama* in *Samayasāra* (*SS*) 36-37, 342 and *Niyamasara* (*NS*) 27, 108.

3

समवाओ पंचण्हं समउत्ति जिणुत्तमेहिं पण्णत्तं ।
सो चेव हवदि लोओ तत्तो अमिओ अलोओ खं ॥

samavāo pañcaṇhaṁ samautti jiṇuttamehiṁ paṇṇattaṁ ।
so ceva havadi lo-o tatto amio alo-o khaṁ ॥

It is said by Jina (the Omniscient Lord) that the proper understanding of the collection or group of five existences (*astikāyas*) constitutes the system of reality; the same is the world. And beyond that is the immeasurable and infinite space, called *āloka*.

Commentary

The system of reality is twofold: *loka* and *aloka*. The *loka* (world or the universe) comprises of the five existences (*astikāyas*). Beyond that is the great and the infinite space (*ananta-ākāśa*), i.e. *āloka*, which is coextensive with pure space.

In this verse, *samaya* is also used in the sense of group, collection or aggregate (*samavāya*) of objects or substances (*padārtha samūha*).

---

1 See *Jainendra Siddhānta Kośa*, vol. 4, p. 328.

Thus, *Pañcāstikāyasāra* 3 states:

> The Omniscient Lord (Jina) has said that defining or determining the group or collection of five *astikāyas* [viz. souls (*jīvas*), inanimate matter (*pudgala*), principle of motion (*dharma*), the principle of rest (*adharma*) and space (*PKS* 4)] with equanimous dispositions (*sama-bhāva pūrvaka*) or their collection (*samavāya*) is called *samaya*: the same is the world (*loka*). And beyond that is the immeasurable infinite space (called *āloka*) (*PKS* 3).

These five entities are called *astikāyas* by the Jainas because they are existences (*asti* or *astitva*, i.e. *asti*ness), [both general (*sāmānya*) and particular (*viśeṣa*) (*PKS* 3 AC)] capable of occupying space and hence are called *kāyas* (corporeals), i.e. tangible or real.

Commentary by Amr̥tacandra

Commenting on this verse, Amr̥tacandra states:

Samaya means a system (collection of connected objects) which is said to be of three kinds: (1) *śabda-samaya* (scripture); (2) *jñāna-samaya* (knowledge); and (3) *artha-samaya* (meaning or sense) thereof.

1. *Sama*, i.e. *madhyastha* (equanimous, indifferent or neutral), meaning that which is not impaired or deformed by attachment and aversion and *vāda* meaning reading or studying of that which is a collection of letters, marks or signs (*padas*) and sentences is *śabda-samaya* (scripture of words).
2. The proper knowledge (*samyak-jñāna*) of the five *astikāyas* on the destruction or the rise (*udaya*) of deluded-view is knowledge scripture (*jñāna-samaya*).
3. That five *astikāyas* – that knowledge through the supportive factor (*nimitta*) of description is *artha-samaya*, i.e. collection of all *padārthas*, is *artha-samaya*.

That *artha-samaya* is of two kinds: (1) as much as the collection of five *astikāyas*, just so much is the world or the universe; and (2) beyond that is immeasurable and infinite space (called *āloka*).

4

जीवा पुग्गलकाया धम्माधम्मा तहेव आयासं ।
अत्थितह्मि य णियदा अणण्णमइया अणुमहंता ॥

jīvā puggalakāyā dhammādhammā taheva āyāsaṁ ।
atthitahmi ya ṇiyadā aṇaṇṇamaiyā aṇumahaṅtā ॥

The five existences (*astikāyas*) are: (1) conscious soul (*jīva*); (2) inanimate, non-conscious matter (*pudgala*); (3) the principle or medium of motion (*dharma*); (4) the principle or medium of rest (*adharma*); and (5) space (*ākāśa*). They are eternal and uncreated and of extensive magnitude.

Commentary

Since the atom or the material point is the unit of space, the spatial point is also called *aṇu* (*atom*). Since the five entities such as *jīvas* mentioned above are capable of occupying space, they are called *kāyas* (corporeals); the word *kāya* implies relation to many spatial points. Existences that can be so related to space are called *astikāyas* by the Omniscient Lords (Jinas).

Dharma and *adharma* are the peculiar principles recognized by Jainas as corporeal. So, these are also *astikāyas*.

Since time (*kāla*) has neither potentially nor really the possibility of such space relations, it is denied the name *astikāya*. It is purely an entity of mono-dimensional series. Though it is not *astikāya*, it does not cease to be a real entity. Therefore, the five *astikāyas* together with time are considered six substances (*dravyas*).

Commentary by Amr̥tacandra

Time does not occupy more than one unit of space, i.e. many spatial points, and does not have extent, i.e. it is not extensive. Therefore, it is not *kāya* (extensive); the units of time (*kālāṇu*) are packed in *loka* (world) like a heap of jewels.

5

जेसिं अत्थिसहाओ गुणेहिं सह पज्जएहिं विविहेहिं।
जे होंति अत्थिकाया णिप्पण्णं जेहिं तेल्लोक्कं ॥

jesiṁ atthisahāo guṇehiṁ saha pajjaehiṁ vivihehiṁ ।
je hoṁti atthikāyā nippaṇṇaṁ jehim tellokakaṁ ॥

Whatever things have the essential nature of manifesting themselves severally through their numerous qualities and more are the *astikāyas.* These fill the three worlds – they being the constituent parts of the world.

Commentary

Since the five *astikāyas* have existence or present reality, i.e. they are real, they are called *asti.* And since they have the space quality of occupying many space points and constitute the world (*loka*), they are called *kāyas. Atikāya*, thus, implies existential nature as well as space quality.

BOOK ONE

Dravya (Substance)

6

ते चेव अत्थिकाया तेक्कालियभावपरिणदा णिच्चा ।
गच्छंति दवियभावं परियट्टणलिंगसंजुत्ता ॥

te ceva atthikāyā tekkāliyabhāvapariṇadā ṇiccā ।
gacchanti daviyabhāvaṁ pariyadṛṇaliṅgasaṁjuttā ॥

The five *astikāyas*, which though transforming themselves diversely in all three times, maintaining their permanence, constitute together with time (*kāla*) form the six substances (*dravyas*). The very term *dravya* signifies *dravyatva* (that which by nature flows towards its modes). And the term "substance" is suggestive of something, that which subsists or persists in spite of the changes, i.e. permanency through change. Thus, the *dravyas* (substances) undergo change and yet maintain their identity.

Commentary by Amṛtacandra

[*Dravya* (substance) is defined as] that which, whilst it does not forsake its innate nature, is characterized by origination (*utpāda*), decay or annihilation (*vyaya*) and permanency (*dhrauvya*), which possesses qualities (*guṇas*) and modes or modifications (*paryāyas*). (*PS* 95; *PKS* 10; *TS* V.30 and 38).

Thus, they (substances) have the three characteristics of coming into existence, ceasing to exist and also remaining permanent in spite of both. Creation, destruction and yet continuity – these are the main characteristics of *dravya.*

The *dravya* is also the single, inalienable substratum (*ādhāra-bhūta*) of qualities (*guṇas*) that are simultaneous (*sahabhāvī*), and modes (*paryāyas*), that are successive (*krama-bhāvī*), i.e. they appear and disappear. Since it (*dravya*) does not give up its deputed (*pratiniyata*) nature, it is only permanent alone.

7

अण्णोण्णं पविसंता दिंता ओगासमण्णमण्णस्स ।
मेलंता वि य णिच्चं सगं सभावं ण विजहंति ॥

aṇṇoṇṇaṁ pavisaṅtā diṅtā ogāsamaṇṇamaṇṇassa ।
melaṅtā vi ya ṇiccaṁ sagaṁ sabhāvaṁ ṇa vijahaṅti ॥

Though the six substances are mutually interpenetrating and accommodating one another and though they commingle and get mixed up in view of occupying the same space, yet they always maintain their own intrinsic nature (*svaka svabhāva*) without losing their respective qualities, general as well as special.

Commentary by Amṛtacandra

Though the six substances are excessively interpenetrating and commingling, they do not give up their own intrinsic nature. Therefore, they are evolvement and yet permanent (*nitya*), i.e. *pariṇāmī nitya.* There is, therefore, no unity or oneness (*ekatva*) between them; though from external, other-referential point of view (*vyavahāra naya*), the Self (*jīva*) and material *karma* or no-*karma* (physical body) are said to have *ekatva* (*SS* 27), they do not grasp the nature of one another.

Commentary by Jayasena

Of the six substances, the soul (*jīva*) and matter (*pudgala*) are active (*sakriya*), i.e. they can move about from one place to another place or evolvent. *Dharma*, *adharma*, *ākāśa* (space) and *kāla* (time) are inactive (*niṣkriya*) *dravyas* as they condition movements, without themselves undergoing change or motion.

Then, describing the characteristics of *saṅkalpa* and *vikalpa*, Jayasena states:

> Having the psychic dispositions of "This is mine" (*mama-edam*), etc. among the conscious or animate, non-conscious or inanimate and mixed animate and inanimate substances, is *saṅkalpa*, while thoughts such as "I am happy", "I am miserable", etc. and the states-of-evolution of joy or delight (*harṣa*) and sadness, detection of despair (*viṣāda*) are the thought-constructions (*vikalpa*).

Describing the nature of existence (*astitva* or *sattā*), Kundakunda states:[1]

8

सत्ता सव्वपयत्था सविस्सरूवा अणंतपज्जाया ।
भंगुप्पादधुवत्ता सप्पडिवक्खा हवदि एक्का ॥

sattā savvapayatthā savissarūvā aṇaṅtapajjāyā ।
bhaṅguppādadhuvattā sappaḍivakkhā havadi ekkā ॥

Sattā (existence): (1) It exists or is present in all substances [i.e. all substances are existent realities]. It is one (as a class). (2) It is of the nature of the inherent essence of all things (*sa-viśvarūpa*). (3) It manifests itself through diverse forms (*sa-pratipakṣa*). (4) It undergoes infinite modifications. (5) It has the triple characteristics of origination, destruction and permanence. (6) It has the antithetical qualities (*sa-pratipakṣa*), i.e. pairs of opposites coexist in the same substance as inalienable parts thereof.

Commentary

According to the Jaina doctrine of *anekānta* (multifaceted nature

1 See Introductory remarks of Amṛtacandra to *Pañcāstikāyasāra* 8.

of reality), the mutually contradictory or opposite properties, characteristics or traits coexist in the same object as inalienable parts thereof. Thus, in the complex reality, the contrary traits of permanence and change, existence and non-existence, one and many, etc. inhere in the same real.

The *pratipakṣa* (opposite) of one substance (say *jīva* (soul)) is another substance (non-soul, *ajīva*) and the *pratipakṣa* of consciousness (*cetana*) is non-consciousness (*acetanā*).

The existence of a jar is affirmed as real with reference to its own intrinsic substance (*sva-dravya*), its own place (*sva-kṣetra*), its own time (*sva-kāla*) and its own state or nature (*sva-rūpa* or *sva-bhāva*), while non-existence of the jar is established with reference to other or extrinsic substances (*para-dravya*), other place (*para-kṣetra*), other time (*para-kāla*) and other state or mode (*para-bhāva* or *para-rūpa*).

Thus, the existence of a thing or its non-existence depends on the four aspects or factors of substance (*dravya*), place (*kṣetra*), time (*kāla*), and state, condition, mode or nature (*bhāva* or *rūpa*) relevant to a situation. Frank van Den Bossche calls these four aspects or determining factors as the "ontological coordinates" of a thing; without these, existence is impossible.

Commentary by Amr̥tacandra

In his commentary on this verse Amr̥tacandra states:

> *Sattā* (existence of reality) is not *niraṅkuśa*, i.e. unrestrained, self-willed, absolute autocratic; but it has its contrary perspective (*sa-pratipakṣa*). It is *sāmānya-viśeṣātmaka*, i.e. it can be viewed with reference to general (*sāmānya*) perspective and also from the particular or special (*viśeṣa*) perspective.

(1) The *pratipakṣa* of existence (*sattā*) is non-existence (*asattā*); (2) the *pratipakṣa* of triple characteristics of origination, decay and permanency (*tri-lakṣaṇa*) is its opposite (*atri-lakṣaṇa*); (3) the *pratipakṣa* of one is many; (4) the *pratipakṣa* of "that which is situated

or existent in all things (*sa-viśvarūpa*) is situated in one object or of being of the nature of one *eka-rūpapana*; and (6) the *pratipakṣa* of having infinite modes or modifications (*ananta siddha māyā*) is having only one mode or modification (*eka-siddhapana*).

The existence (*sattā*) is twofold: (1) *mahā-sattā* (great, mighty, generally existent, reality); and (2) *avāntara sattā* (intermediate *sattā*). *Mahā-sattā* is that which is pervasive in all objects indicative of existence of common nature (*sādr̥śya astitva*). *Avāntara sattā* is indicative of existence-of-own-nature (*svarūpa-astitva*).

Description of the nature of *sattā* is dependent on two viewpoints or the two ways (*prarūpaṇa*) devised to deal with general or universal (*sāmānya*) and particular or specific (*viśeṣa*).

9

दवियदि गच्छदि ताइं ताइं सब्भाव पज्जयाइं जं।
दवियं तं भण्णंति अण्णण्णभूदं तु सत्तादो ॥

daviyadi gacchadi tāiṁ taiṁ sabhāva pajjayāiṁ jaṁ ।
daviyaṁ taṁ bhaṇṇaṁti aṇṇaṇṇabhūdaṁ tu sattādo ॥

That which by nature flows towards its modes (*dravyatva*) and yet maintains its identity (through it several qualities and modifications (*siddha*), and is not other than or inalienable (*ananyabhūta*) from existence (*sattā*) is called *dravya* (substance).

Commentary by A. Chakravarti

In this verse, *sattā* (substance) is distinguished from *dravya*, which means that which flows or changes. While changing through its different qualities and modifications, its essential nature persists. This kind of progressive development is associated with *dravya*. But such a development is also the characteristic of substance. Hence, according to Jaina doctrine, *dravya* is not entirely different from *sattā* (substance). Therefore, the opposite qualities mentioned above with reference to *sattā* are also applicable to *dravya*, which is not different in meaning.

According to this view, there is no unchanging substance (*sattā*) in Jainism. Such adamantine existence cannot be identified with *dravya* which is extremely volatile. Here also, the similarity between the Hegelian concept of a "thing" and the Jaina concept of *dravya* is noticeable. *Sattā* is not a "thing-in-itself" behind *dravya*. *Sattā* and *dravya* are one and the same as mentioned by Georg W.H. Hegel. A "thing-in-itself" and experience are not absolutely distinct. *Dravya* refers to facts of experience, whereas *sattā* refers to existence or reality. One may be abstracted from the other, but it is not in reality, different from the other.

10

दव्वं सल्लक्खणियं उप्पादव्वयधुवत्तसंजुत्तं ।
गुणपज्जयासयं वा जं तं भण्णंति सव्वण्हू ॥

davyaṁ sallakkhaṇiyaṁ uppādavvayadhuvattasaṁjuttaṁ ।
guṇapajjayāsayaṁ vā jaṁ taṁ bhaṇṇaṁti savvaṇhū ॥

The existent reality (*sat*) is the characteristic of *dravya* (substance) (*TS* V.29). *Dravya* is characterized by origination, decay or destruction and permanence. It is also the substratum of its qualities (*guṇas*) and modifications (*siddhas*).

Commentary

In this verse, three characteristics of *dravya* are described. These three characteristics express the same essential nature of *dravya* in three different ways.

The characteristic that *dravya* is the existent reality (*sat*) refutes the Buddhist belief of momentary existence (*Kṣaṇikavāda*). That *dravya* is characterized by origination, decay or destruction, which accounts for change and yet remains permanent, differs from Vedāntic monism, and Sāṁkhya and Naiyāyika systems of philosophy which emphasize absolutely unchangeable permanency (*kūṭastha nitya*). Kundakunda has criticized the Sāṁkhya system in his *Samayasāra* 116-125. Reality cannot be absolutely eternal and unchangeable. If there is no modification or change, there can be no right action or wrong action, merit or demerit, bondage, transmigration or liberation (*akalaṅka*). The one-sided emphasis either on permanency or change is rejected by Jaina thinkers.

The Jaina thinkers' doctrine of *anekānta* (multifaceted nature of reality) maintains that reality has multifarious aspects and that a complete comprehension of that must necessarily take into consideration all the different aspects through which reality manifests itself.

11

उप्पत्तीव विणासी द्व्वस्स य णत्थि अत्थि सब्भावो ।
वयं उप्पादंधुवत्तं करंति तस्सेव पज्जायाः ॥

uppattīva viṇāsī davvassa ya ṇatthi atthi sabbhāvo ।
vayaṁ uppādaṁdhuvattaṁ karaṅti tasseva pajjāyāḥ ॥

In reality, *dravya* can neither be created nor destroyed; it has only permanent substantiality. It is only the modes (*siddhas*) that appear and disappear. Thus, it is through its modes that *dravya* obtains the triple characteristics of appearance, disappearance and permanency.

Commentary by A. Chakravarti

According to *dravyārthika naya* (substance point of view), e.g. gold can neither be created nor destroyed. It exists and that is all. But the ornaments and other things made of gold are the *paryāyas* (modes). These may change; one ornament may be melted and a new one be made. The disappearance of the previous ornament is *vyaya* (decay) and the appearance of the new ornament is *utpāda* (origination) and yet all the while there is the same gold *dhruva* (permanent). *Dravya* then has both the qualities, permanence and change, it is permanent as *dravya* and changing as *paryāya* (modes).

12

पज्जयविजुदं दव्वं दव्वविजुत्ता य पज्जया नत्थि।
दोण्हं अणण्णभूदं भावं समणा परूविंति॥

pajjayavijudaṁ davvaṁ davvavijuttā ya pajjayā natthi ।
doṇhaṁ aṇaṇṇabhūdaṁ bhāvaṁ samaṇā parūviṁti ॥

There is neither substance without the mode, nor mode (*siddha*) without the substance. One cannot be without the other as they are inalienable (*ananyabhūta*), so say the saints (*śramaṇas*).

Commentary

Though there is a difference between the substance and its modes with respect to the designation (*saṁjñā*) and utility or *prayojana* (purpose), the one cannot exist apart from the other, as the two are inalienable states (*ananyabhūta bhāva*). There can be no ornament without the gold and only the gold is transformed into some ornament.

13

दव्वेण विणा ण गुणा गुणेहिं दव्वं विणा ण संभवदि।
अव्वदिरित्तो भावो दव्वगुणाणं हवदि तह्मा॥

davveṇa viṇā ṇa guṇā guṇehiṁ davvaṁ viṇā ṇa saṁbhavadi ।
avvadiritto bhāvo davvaguṇāṇaṁ havadi tahmā ॥

There is neither quality without substance nor substance without quality. The one cannot exist apart from the other as there is *avyatirikta bhāva* (inseparable, integral or inalienable). The two are compatible to one another.

Commentary

The substance and quality are different with respect to designation appellation, yet they are not really different in nature as both refer to the same real fact.

14

सिय अत्थि णत्थि उहयं अव्वत्तव्वं पुणो य तत्तिदयं ।
दव्वं खु सत्तभंगं आदेसवसेण संभवदि ॥

siya atthi ṇatthi uhayaṁ avvattavvaṁ puṇo ya tattidayaṁ ।
davvaṁ khu sattabhaṅgaṁ ādesavaseṇa saṁbhavadi ॥

Because of the directive, injection or mandate of the seven modes of thought or possible predications (*sapta-bhaṅga*), the *dravya* is viewed as (1) *syāt*, i.e. in a certain context, in a certain sense, in some respect or from a particular point of view, it certainly is; (2) *syāt* it is not; (3) *syāt* it is both (is and is not); (4) *syāt* it is indescribable or inexpressible (*avaktavya*); (5) *syāt* it is and is inexpressible; (6) *syāt* it is not and is inexpressible; and (7) *syāt* it is and it is not and is inexpressible.

Commentary

The word *syāt* does not mean probability, perhaps (as Chakravartinayanara has translated), doubt or indefiniteness, as many people mistakenly believe. It makes a definitive assertion (often it is accompanied by the term *eva* (certainly), which confers definiteness) and emphasizes a conditional or relative character of a statement.

Thus, a jar certainly exists from its own (*sva*) substance (*dravya*), its own place (*kṣetra*), its own time (*kāla*), and its own state, mode or condition (*bhāva* or *rūpa*), i.e. it exists from its own quartet or quarternary (*sva-catuṣṭaya*). But it certainly does not exist with reference to the other substance (*para-dravya*), other place (*para-*

kṣetra), other time (*para-kāla*) and other state or condition (*para-bhāva*), i.e. the other four aspects (*para-catuṣṭaya*).

15

भावस्स णत्थि णासो णत्थि अभावस्स चेव उप्पादो।
गुणपज्जयेसु भावा उप्पादवए पकुव्वंति ॥

bhāvassa ṇatthi ṇāso ṇatthi abhāvassa ceva uppādo ।
guṇapajjayesu bhāvā uppādavae pakuvvaṁti ॥

There is no destruction of existent realities, i.e. of things that do exist, nor can there be origination of things out of nothing. Coming into existence and ceasing to exist takes place in things that have attributes and modes.

Commentary by Amr̥tacandra

There can be neither destruction of existing substances (*dravyas*) nor origination of non-existing ones. Concepts of origination and destruction are applicable to a substance only because of its qualities or attributes (*guṇas*) and modes (*siddhas*). In this verse *dravya* is shown to be permanent from the substance point of view (*dravyārthika naya*) and changing from modifications point of view (*paryāyārthika naya*).

In this verse the appearance or origination (*utpāda*) of *asat* (not-existent) and destruction of *sat* is denied. In other words, without destruction of *sat*, i.e. existent reality and without the origination of *asat*, destruction and origination take place in qualities or attributes (*guṇas*), and modes or modifications (*paryāyas*).

For instance, the atoms of gold, that constitute the substance gold, are subject to neither origination nor destruction. But there may be appearance and disappearance in the different forms and modifications of gold: the original form may be lost and a new form may be assumed. One ornament may be destroyed and another created.

16

भावा जीवादीया जीवगुणा चेदणा य उवओगो।
सुरणरणारयतिरिया जीवस्स य पज्जया बहुगा॥

bhāvā jīvādīyā jīvaguṇā cedaṇā ya uvaogo ।
suraṇaraṇārayatiriyā jīvassa ya pajjayā bahugā ॥

Jīva and other *dravyas* are reals or existent realities. The qualities of *jīva* and consciousness (*cetanā*), and conscious attentiveness are manifestation or function of consciousness, i.e. *upayoga* (having intuition and knowledge). The soul manifests in various forms or modes (*siddhas*) of a celestial being, a human being, denizens of hell, and plant and animal life.

Commentary

Although several features such as existence, substantiality/ functionality, changeability and knowability are shared by *jīva* in common with physical objects or matter, consciousness remains the defining characteristic, which distinguishes *jīva* (soul) from inert, inanimate matter. The number of bio-energies (physical *prāṇas*) may be more or less in different living organisms and the psychic potentialities may also differ from one living being to another. Consciousness, however, is the essential ground condition, feature or characteristic of life, which remains throughout the entire journey of life and without which the existence of a living organism is inconceivable.

In Jaina texts, the conscious entity is described as *svaparaprakāśam* (Kundakunda, *Niyamasāra* 161) or *svaparāvabhāsakam* (Samantabhadra, *Svayambhū-stotra* 63), i.e. that which illumines itself as well as other objects. Consciousness is, thus, an awareness of one's own Self as well as awareness of the objects of the world, i.e. being aware of something. According to Sri Aurobindo also, consciousness is the power of awareness of the Self and things.

Moreover, conscious attentiveness (*upayoga*) represents the most significant and special function of the conscious entity (*cetana dravya* or *jīva*). We often pass through a street without hearing or seeing certain things because our consciousness was not attentive to them; we were absorbed otherwise. Therefore, besides consciousness, this attentiveness of consciousness (*upayoga*) is also considered another definition or the distinguishing characteristic (*lakṣaṇa*) of *jīva* (soul), which finds mention in Jaina texts (Umāsvāmi or Umāsvāti, *Tattvārthasūtra* II.8). It must, however, be noted that *upayoga* (psychic attention) is not the same as consciousness.

Upayoga (activity or function of consciousness) is different and distinct (*vibhakta*) from consciousness (*cetanā*) with reference to nomenclature or name (*saṁjñā*), number (*saṅkhyā*), characteristics (*lakṣaṇa*) and purpose (*prayojana*). The names of *cetanā* and *upayoga* are different (*saṁjñā*). While *cetanā* is one, *upayoga* is threefold, viz. *darśana*, *jñāna* and *caritra* (difference in number (*saṅkhyā*)). Of the five thought activities (*bhāvas*) of the Self (*TS* II.1), the first four, viz. the one relating to coming into operation a rise (*udaya*) of *karmas* (*audāyika*), the subsidential of *karma* (*aupaśamika*), destruction of *karmas* (*kṣāyika*) and the mixed process of destruction and subsidence of *karmas* (*kṣayopaśamika*) are *karma*-referential. The last one, the *pariṇāmika bhāva* is non-relational to or independent of *karmas* (*karma-nirapekṣa*) and is uncontaminated and unconditioned by *karmas*. It is not casually connected with the transmigratory existences (*saṁsāra*) or liberation (*mokṣa*); it is *niṣkriya* (inactive) *bhāva* (see *PKS* 56 JS). *Upayoga*, on the other hand, is the activity of consciousness (difference in *lakṣaṇa*). Moreover, while the real nature of the Self is pure in *karma-nirapekṣa* sense, *upayoga*, when impure, is the cause of conjunction with the other substance (*PS* 156 AC), accounting for its impurity (difference in *prayojana* (purpose)).

Conscious attentiveness or the function of consciousness (*upayoga*) can be auspicious (*śubha*), virtuous or meritorious (*puṇya*) or inauspicious (*aśubha*) or sin (*pāpa*) (*PS* 155), and as such accounts for the accumulation of *karmas* (*PS* 156).

Jīva is declared to be a consciousness entity (*cheda*) (*SS* 297). As such, it is identified with consciousness. *Upayoga* is considered to be the spatial of a particular (*viśeṣa*) property, the characteristic or modification of consciousness (*PKS* 27).

Akalaṅka defines *upayoga* as *caitanya anuvidhāyī pariṇāma*, i.e. the state-of-evolution of *jīva* in obedience to consciousness. Consciousness is verily consciousness alone (*Samayasāra Kalaśa* 70-89). It is the very nature of the soul and the substratum of its qualities and modifications, and all forms of conscious activity, pure and impure.

17

मणुसत्तणेण णट्ठो देही देवो हवेदि इदरो वा ।
उभयत्त जीवभावो ण णस्सदि ण जायदे अण्णो ॥

maṇusattaṇeṇa naṭṭho dehī devo havedi idaro vā ।
ubhayatta jīvabhāvo ṇa ṇassadi ṇa jāyade aṇṇo ॥

If a soul departs from the state of a human being, it becomes the celestial being (*deva*) or some other living being. In either case (during death or birth), its intrinsic nature (*jīva bhāva*) is not destroyed and other (*anya*) *jīva bhāva* is not created.

Commentary

The change from a human mode to a celestial being takes place in the mode (*siddha*); the soul substance (*dravya*) remains the same, i.e. maintains its nature and identity.

The Jaina view rejects as untenable the Buddhist concept of momentariness (Kṣaṇikavāda), which maintains that there is a different Self at every moment. On the other hand, Advaita Vedānta maintains that the Self is absolutely permanent and unchanging and that all changes are illusory. This view is also rejected by the Jainas.

18

सो चेव जादि मरणं जादि ण णट्ठो ण चेव उप्पण्णो ।
उप्पण्णो य विणट्ठी देवो मणुसूत्तिपज्जाओ ॥

so ceva jādi maraṇaṁ jādi ṇa ṇaṭṭho ṇa ceva uppaṇṇo ।
uppaṇṇo ya viṇaṭṭhī devo maṇusūttipajjāo ॥

Though the soul experiences both birth and death, it is neither really destroyed nor created. The destruction of the human mode and the appearance of the birth of a celestial being are only in the modes (*siddhas*) of the soul.

Commentary by Jayasena

From the substance point of view (*dravyārthika naya*), the soul is permanent and the indestructible essence, and hence there is neither birth nor death of the soul. It is only from the modification point of view (*paryāyārthika naya*) that origin and decay or destruction take place.

Thus, from the *dravyārthika* and *paryāyārthika* points of view, there is both permanence and impermanence in only one substance.

19

एवं सदो विणासो असदो जीवस्स णत्थि उप्पादो ।
तावदिओ जीवाणं देवो मणुसोत्ति गदिणामो ॥

evaṁ sado viṇāso asado jīvassa ṇatthi uppādo ।
tāvadio jīvāṇaṁ devo maṇusotti gadiṇāmo ॥

Thus, the existent reality or thing is neither liable to death nor destruction, nor a non-existent thing can come into being. It is said that the human mode has undergone death and a celestial being is born. The reason for that is that all this takes place due to *nāma-karma* in respect of *gati* (condition of existence).

Commentary

There are four classes/conditions of existence (*gati*): (1) human being (*manuṣya*); (2) celestial being (*deva*); (3) the denizens of hell (*naraka*); and (4) animal or plant life (*tiryañc gati*). The duration of each state or condition of existence is determined by its own *nāma-karma*.

20

णाणावरणादीया भावा जीवेण सुट्ठुअणुबद्धा ।
तेसिमभावं किच्चा अभूदपुव्वो हवदि सिद्धो ॥

ṇāṇāvaraṇādīyā bhāvā jīveṇa suṭṭhuaṇubaddhā ।
tesimabhāvaṁ kiccā abhūdapuvvo havadi siddho ॥

Due to impure psychic dispositions (*bhāva karmas*), knowledge-obscuring, etc. physical *karmas* (*dravya karmas*) are connected with *jīva* (Self). But by the absence of the bondage of *karmas* through his exertion, the Self attains the unprecedented (*abhūtapūrva*) liberated state of *siddha*, which is the mode of *jīva* (soul).

Commentary by A. Chakravarti

Though gold may be different from brass, it is mixed with it. Similarly, the *jīva* pure in itself gets bound by *karmas*. The initial condition is its own impure heart called *bhāva karma*. Owing to this psychological state, *dravya karmas*, or the different kinds of kārmic matter, are attracted and get deposited on the *jīva*, thereby shrouding its light and glory.

When the *jīva* is so bound by *karmas*, it undergoes a series of modifications. When it is eventually able to realize its true nature, it may liberate itself from kārmic shackles and attain *nirvāṇa*. When this stage is reached, there is no apprehension of coming back to *saṁsāra*. Jainas do not assume the doctrine of the *fall of man*. Every *jīva* is initially a *karma-jīva* and *mokṣa* is a unique state to be attained anew and for the first time. The state of nature is not a state of freedom. It is a state of bondage. *Jīva* finds itself in chains, and by its own efforts it attains freedom.

21

एवं भावमभावं भावाभावं अभावभावं च ।
गुणपज्जयेहिं सहिदो संसरमाणो कुणदि जीवो ॥

evaṁ bhāvamabhāvaṁ bhāvābhāvaṁ abhāvabhāvaṁ ca ।
guṇapajjayehiṁ sahido saṁsaramāṇo kuṇadi jīvo ॥

Thus, *jīva* (soul) with its attributes and modes, wandering in the transmigratory existences of births and deaths (*saṁsāra*), may lose its particular form and assume a new one. Again, this (new) form may be lost and the original (the previous form or mode) may be regained.

Commentary

Changes take place in forms or modes (*paryāya*), which appear and disappear, but the Self maintains his identity through these changes.

22

जीवा पुग्गलकाया आयासं अत्थिकाइया सेसा ।
अमया अत्थित्तमया कारणभूदा हि लोगस्स ॥

jīvā puggalakāyā āyāsaṁ atthikāiyā sesā ।
amayā atthittamayā kāraṇabhūdā hi logassa ॥

The souls, material bodies, space together with the principle of motion (*dharma*) and the principle or medium of rest (*adharma*) are uncreated existences, that constitute the world. These are the *astikāyas*.

Commentary

There are an infinite number of souls and an infinite number of matter particles. The five *astikāyas* are uncreated and eternal. The Jainas recognize the reality of space and of *dharma* and *adharma*.

23

सब्भाव सभावाणं जीवाणं तह य पोग्गलाणं च ।
परियट्टणसंभूदो कालो णियमेण पण्णत्तो ॥

sabbhāva sabhāvāṇaṁ jīvāṇaṁ taha ya poggalāṇaṁ ca ।
pariyadr̥ṇasaṁbhūdo kālo ṇiyameṇa paṇṇatto ॥

That on account of which these existences, the souls, material particles, *dharma* and *adharma*, undergo changes is called real-time (*kāla-dravya*).

Commentary by A. Chakravarti

Time is of two kinds, viz. real or absolute and relative. Birth, growth and decay of things are possible only because of the former absolute time. The relative time is made up of conventional periods measured by units based upon these changes. This is merely an aspect of real time which is constituted by instants. This is unconditional and absolute.

One who accepts the reality of the concrete world and its changes, must also admit the reality of time. If change is real, time must also be real. And lastly, the absolute or real time is constituted of instants which correspond to the points of space and particles of matter.

In fact, these three classes of elements, points, instants and particles have important functions in the Jaina system of philosophy. It is also worth noting that modern realists led by the mathematical philosophers admit the doctrine that space is real and is made up of points. Time is real and is made up of instants or moments, and the physical world is real and is made up of particles. All these doctrines form the fundamental concepts of the Jaina world-view.

24

ववगदपणवण्णरसो ववगददोगंधअट्ठफासो य ।
अगुरुलहुगो अमुत्तो वट्टणलक्खो य कालोत्ति ॥

vavagadapaṇavaṇṇaraso vavagadadogaṁdha-aṭṭhaphāso ya ।
agurulahugo amutto vaḍṛṇalakkho ya kālotti ॥

What is without the five colours in the five days, without the two smells and eight kinds of touch (*sparśa*); and which is characterized by *agurulaghu*, i.e. (1) having constancy or individuality, the quality by which a substance or attribute does not become another, and by which none of the attributes of a substance can never fall off from it; (2) absence of family-determining *karma*; (3) neither too heavy nor too light; and (4) increase and decrease in various ways in the modifications of all substances. It is a common feature of all *dravyas* and the natural feature found in a mass and also in an atom.

Commentary by A. Chakravarti

Time has meaning for and existence in the world. It is made up of instants (*kālāṇus*). The instants have neither colour nor taste nor touch. The qualities associated with physical objects cannot be applied to it. They constitute a "continuous compact series". The time series formed by instants is one-dimensional in the language of the mathematicians; that is why time is denied *kāyatva* by Jaina philosphers. Time, which is so constituted by instants, is called *mukhya-kāla* (absolute time). It is also described by the following terms: *paramārtha-kāla* (great and unconditioned time); *niścaya-kāla* (true time) and *dravya-kāla* (real time). It is one of the six *dravyas* (reals).

25

समओ णिमिसो कट्ठा कला य णाली तदो दिवारत्ती ।
मासोदुअयणंसंवच्छरोत्ति कालो परायत्तो ॥

samao ṇimiso kaṭṭhā kalā ya ṇālī tado divāratti ।
māsoduayaṇaṁsaṁvaccharotti kāla parāyatto ॥

Sāmya, nimiṣa, kāṣṭhā, kāla, nālī, then (*muhūrta*), day and night, month, season, *ayana* and *saṁvatsara,* and other periods of time – all these are called external, other-referential. They are all *vyavahāra* (conventional) time. They are dependent on or determined by other (*parayattaḥ* or *para-āśrita*).

Commentary by A. Chakravarti

Sāmya (moment) which is the manifestation of infinite number of ultimate instants.

Nimiṣa which is made up of innumerable such moments.

Kāṣṭhā which is equal to fifteen *nimiṣas.*

Kāla which is equal to thirty *kāṣṭhās.*

Nālī which is a little over twenty *kālas.*

Muhūrta which is equal to two *nālīs.*

Day and night, equal to thirty *muhūrtas.*

Month which is equal to thirty days.

Season which is equal to two months.

Ayana half an year equal to three seasons.

Saṁvatsara year it is equal to two *ayanas.*

The different periods enumerated in this verse are the conventional periods measured by different units. These artificial distinctions are introduced into the continuous time series of real time. The units which determine the conventional distinctions are generally the changes in the objects of the physical world or of the organic world. Winking of the eyelid, the diurnal motion of the sun across the heavens (which is only apparent according to modern astronomy) and the motion of the moon around the earth are some of the important changes of physical objects which have been traditionally associated with a conventional period of measurements. Because of these conventional distinctions, relative time is said to be *parayattā* (dependent on other things), but still it is merely a form of real time which is an unconditioned *dravya*. *Vyavahāra-kāla* is said to be conditioned *in a way*. It is called *kathācita parayattā*, which is in some respect conditioned by the other.

26

णत्थि चिरं वा खिप्पं मत्तारहिदं तु सा वि खलु मत्ता।
पुद्गलदव्वेण विणा तह्मा कालो पडुच्चभवो॥

ṇatthi ciraṁ vā khippaṁ mattārahidaṁ tu sā vi khalu mattā ।
pudgaladavveṇa viṇā tahmā kālo paḍuccabhavo ॥

Duration of time, either long or short, is impossible apart from the standard of measurement. The standard of measurement also has no meaning apart from material objects. Hence, conventional or relative time is brought about by extraneous conditions.

Commentary by A. Chakravarti

Vyavahāra-kāla (conventional time) consists of periods of duration which may be either short or long. This distinction of length of periods is not intrinsic. It is entirely due to external conditions such as the moments of the physical bodies. Therefore, though time as such is unconditioned, *vyavahāra-kāla* is certainly conditioned by alien objects. Thus, time as such is the instrument of change or *pariṇāma* (state-of-evolution) in the five existences of the world, while itself has the *vyāvahārika paryāyas* (modes) measured by the changes in the other objects.

CHAPTER ONE

Jīva (Self)

27

जीवोत्ति हवदि चेदा उपओगविसेसिदो पहूकत्ता ।
भोत्ता य देहमत्तो ण हि मुत्तो कम्मसंजुत्तो ॥

jīvotti havadi cedā upaogavisesido pahūkattā ।
bhottā ya dehamatto ṇa hi mutto kammasaṁjutto ॥

A comprehensive description of the characteristics of the individual Self (*jīva*) is given in this verse.

Commentary

The characteristics of the *jīva* are as follows:

1. That which lives, will live and has lived previously with *dravya prāṇas* (physical vitalities, bio-energies), viz. five senses, three energies of mind, speech and body, life-span and respiration, is *jīva* (*PKS* 30) (from external, other-referential point of view (*vyavahāra naya*), while from internal, self-referential point of view (*niścaya naya*), the Self has *jīvatva* (essence of life) due to possession of conscious vital force (*bhāva prāṇa*).

   The mundane soul is living due to possessing both *dravya prāṇas* and *bhāva prāṇas* while the liberated soul lives only by having *bhāva prāṇa* (*PKS* 30 AC). In the mundane (*saṁsāra*) condition, the Self (*ātman*) is *sopādhi*, i.e. having *upādhi* (limitations of kārmic adjuncts or encumberances) and also it is of the nature of *nirupādhi* meaning the innate, pure nature of the Self devoid of or free from *upādhi* (*PKS* 27 AC). It is *nirupādhi* from the pure *niścaya* point of view and *sopādhi* from the impure *niścaya* point of view (*PKS* 27 JS).

2. The *jīva* is the conscious entity endowed with consciousness, which is its essence and internal life source.
3. It has *upayoga* (conscious attentiveness, psychic attention, function or manifestation of consciousness), which is the particular property or characteristic of consciousness which distinguishes living from non-living. *Upayoga* is of two kinds, viz. intuition and knowledge. It is also said to be of three kinds with the addition of volition or conduct (*SS* 90, 94-98).
4. The Self is the architect or lord (*prabhu*) of his own destiny, who through his own efforts obtains full freedom bearing all the while full moral responsibility for conduct. He is responsible for his rise and fall. He is neither helpless in the hands of the external environment nor a slave to his internal, subconscious compulsions, instinctive urges and impulsive drives or the neuro-biological processes or brain states.

Jainism lays great emphasis on self-restraint and self-discipline, which are the foundation of the Jaina code of moral conduct and ensure peace, happiness and freedom of the individual as well as spiritual well-being.

There is now increasing awareness in the world that "self-imposed limitations can help to free us". A renowned American heart specialist points out: "What appears like self-restraint can be self-empowerment". Ultimately, it is a choice between "true freedom or slave to our compulsions."[1] The characteristic of *prabhu*, thus, accounts for freedom of action or free will, as it is called in modern psychology. The

[1] D. Ornish, 2001, *Eat More, Weigh Less,* pp. 73-74, New York: Harper Collins Quill.

existence of "conscious free will", that can block or "veto" the process initiated unconsciously, resulting in no motor act is conclusively proved by the experiments of neurologist Benjamin Libet.[2] "Conscious will", he adds, "can stop or veto the process so that no act occurs". In such a case, free will could control the outcome. "That fits with our feeling that we can control ourselves, something that ethical systems urge us to do."[3]

5. Unlike the *puruṣa* of Sāṁkhya, who is a passive spectator, the Jaina Self is an active causal agent (*kartā*) of his own actions. He is responsible for all his actions, good and bad, pure and impure, physical and mental. Being evolvent by nature, the soul accepts the character of its state-of-evolution. When the soul evolves through good or evil, it becomes good or evil. When it evolves through the pure, it becomes pure (*PS* 9).

6. *Jīva* is the enjoyer or the experiencer (*bhoktā*) of the fruits of his own actions. If he is not held responsible for the consequences of his actions, there will be no need of moral and spiritual discipline and the quest for liberation.

   It is quite untenable to hold that the responsibility of doing actions lies with some person and the obligation of experiencing the consequences thereof lies with some other individual. Unlike the Sāṁkhya viewpoint which holds that the soul is an inactive entity but is an enjoyer in an indirect manner, i.e. through intellect which is an evolute of unconscious *prakṛti*, Jainism affirms that activity and enjoyment, i.e. the feelings of pleasure and pain, are the functions of the same conscious entity.

   The statement that *jīva* is the *bhoktā* of his own actions also

---

[2] Benjamin Libet, 2003, "Can Conscious Experiences Affect Brain Acitivity?", *Journal of Consicousness Studies*, 10: 24-25.

[3] Benjamin Libet, 2004, *Mind Time: The Temporal Factor in Consciousness*, p. 197, Cambridge, MA: Harvard University Press.

refutes the Buddhist view that the soul that acts and that which enjoys are two different entities.

7. Being embodied, *jīva* (Self) is said to be of the same size or dimensions as the body, large or small, in which it resides with the result that sensation (*saṁvedanā*) is felt in all parts of the body. It resembles a lamp which illumines the whole of the space enclosed in a small or big room in which it happens to be placed. The Jaina concept of *jīva* (consciousness), being equal in extent to its body ensures that the subjective experience of a person is his own personal experience and not of any third person. This rejects the theory that the soul is a minute atom situated inside the particular organ of the body, such as the heart or the brain.

   The Jaina doctrine that the soul is coextensive with the body appears to anticipate modern somatic philosophers such as Maurice Jean Jacques Merleau-Ponty, Gabriel Marcel and Michael Polanyi. However, unlike the Jaina view that the body can only have physical qualities, somatic philosophers maintain that the body comprises emotions, which are contrary to common experience. However, reflection shows that feelings are not unconscious states. The Sāṁkhya position is also untenable because in its view "unconscious *prakṛti* carries both psychological and physical qualities", while the *puruṣa* (Self) remains "totally isolated"[4] because its involvement in matter is illusory.[5]

8. *Jīva* is non-corporeal (*amūrta*). Being a spiritual entity, the innate pure nature of the Self, it has no corporeal form. It is *amūrta* in spite of its being associated with kārmic matter.

---

4 N.F. Gier, 1996, "Jaina Superhumanism and Gnostic Titanism", *The Scottish Journal of Religous Studies* (Spring and Autumn): 152-153.

5 N.F. Gier, 1995, "Ahimsa, the Self and Postmodernism: Jain Vedantist and Buddhist Perspectives", *International Philosophical Quarterly*, 35(1), Issue no. 137: 72-73.

9. In its corporeal existence, *jīva* is associated with material *karmas* (*karma-saṁyukta*). The conscious entity's involvement with *karma* is merely an "association" (*ekakṣetrāvagaha*, literally, occupying the same space or locus), but there is no direct or actual contact between them.[6]

In Jaina thought, "action" (*karma*) does not mean only those things which are physically and ostensibly done but also includes crystallized effects of past activities of energies. These energies are potentialities, latent in the organism which are lying in our deeper level of subconscious, much below the threshold of conscious awareness.

A great portion of our personality, thus, lies hidden in the depth of the subconscious or unconscious (*ajñāta*) (Umāsvāmi or Umāsvati, *Tattvārthasūtra* VI.6) and what we are aware of (*jñāta*) and to which we pay conscious attention is but a fractional aspect thereof. These *karmas* naturally affect our psychic dispositions and thought activities. They account not only for differences in the physical forms, features and biological structures, including possession of a few or more bio-energies and *paryāpti*, i.e. complete or incomplete development of its various capacities, but also for differences in mental and psychological make-up.

The *karmas*, though primarily material in nature, are not "purely material", as Oriental scholar H. Jacobi mistakenly believed and for that matter assumed that the soul is of some subtle matter thus making the combination between the two possible.[7] *Karmas* are also not mere traces, or psychic impressions or tendencies (*saṁskāras*) as other systems of Indian philosophy maintain.

Jaina thought divides *karmas* into physical *karma* and psychic

---

6 For a detailed discussion of the doctrine of *karmas*, see Jagdish Prasad Jain, 2005, *Fundamentals of Jainism*, pp. 242-264, New Delhi: Radiant.

7 A. Chakravartinayanar, 1975, *Panchastikayasara: The Building of the Cosmos*, p. xxvii, Varanasi: Bhartiya Jnanapeeth.

karma. The physical *karma* is material in nature and can be equated with neuro-biological processes, brain states or neuronal firings and can be said to be the neural correlates of consciousness, while psychic *karma* comprises psychic dispositions and mental states. The physical *karma* has the capacity or the power to influence, affect and condition the mental states, while the mental state, in turn, acts as an indirect, extrinsic, supportive factor or auxiliary cause (*nimitta kāraṇa*) in the mutation of matter into material *karma* (there is difference between matter and material *karma*). The physical *karma* and mental states are mutually related as cause and effect, each of the other, and interact with one another, each acting as an indirect, extrinsic, auxiliary or a subsidiary cause (*nimitta karaṇa*) of the other.

28

कम्ममलविप्पमुक्को उढ्ढं लोगस्स अंतमधिगंता।
सो सव्वणाणदरसी लहदि सुहमणिंदियमणंतं॥

kammamalavippamukko uḍhḍhaṁ logassa aṁtamadhigaṁtā ।
so savvaṇāṇadarasī lahadi suhamaṇiṁdiyamaṇaṁtaṁ ॥

The Self (*ātmā*), which is free from the defects of *karmas*, gets to the highest point of the universe [i.e. the summit (place), which is the abode of liberated souls (*siddhas*)] knows all and perceives all, i.e. it is omniscient and obtains the non-sensual, infinite bliss or happiness that is everlasting.

Commentary by A. Chakravarti

Of the nine attributes, the attribute of lordship is taken up first for explanation. It is known by experience and by the study of the scriptures that the soul is the Lord of his own attributes and states of existence. Impurity of the heart or false faith may lead him into *saṁsāra*. The Lordship should not therefore be interpreted only with the reference to the fall from a high state. Even when he is on the right path, the path that takes him to the never-before-experienced bliss, everlasting, he is the lord of his own destiny. He is not to be imagined merely as a driftwood carried up and down by the waves in the ocean of *saṁsāra*. That would not make him helpless in the hands of extrinsic kārmic forces. He is the creator; the *karmas* themselves owe their existence to his will. Hence, he is his own Lord whether he walks the path of righteousness or chooses the other one. *Ātmā* is his own maker and Lord.

Jaina philosophy is peculiarly associated with its own cosmogony; heavens and hells are arranged in an order. The soul that gets liberation quits the place, rises up to the summit of *loka* where he retains his abode. This dogma is referred to in this verse. As the natural and inevitable result of self-realization, the true Lordship of the soul consists in inheriting the abode of bliss at the summit of the world.

29

जादो सयं स चेदा सवण्हू सव्वलोगदरसी य।
पप्पोदि सुहमणन्तं अव्वाबाधं सगममुत्तं॥

jādo sayaṁ sa cedā savaṇhu savvalogadarasī ya ।
pappodi suhamaṇantaṁ avvābādhaṁ sagamamuttaṁ ॥

That conscious being becoming omniscient and all perceiving through his own efforts, obtains unobstructive infinite happiness or bliss, which transcends sense experience and which is free from imperfection, and is spiritual and incorporeal (*amūrta*).

Commentary

In *Pañcāstikāyasāra* 27, it was stated that the Self is the Lord (Prabhu) of its own destiny; in this verse, the author points out infinite bliss is an acquisition of the Self through his own efforts, without any extraneous help.

30

पाणेहिं चदुहिं जीवदि जीवस्सदि जो हु जीविदो पुव्वं।
सो जीवो पाणा पुण बलमिदियमाउ उस्सासो॥

pāṇehiṁ caduhiṁ jīvadi jīvassadi jo hu jīvido puvvaṁ ।
so jīvo pāṇā puṇa balamidiyamāu ussāso ॥

The Self, which has four bio-energies or physical vitalities (*dravya prāṇas*), is living at present, will continue to live in the future and was living in the past, that same is *jīva*. The *prāṇas* are: (1) three energies or strength (*bala*) of mind, speech and body; (2) five senses (of touch, taste, smell, sound and vision); (3) the duration of life; and (4) respiration. If we take three energies and five senses separately, it makes a total of ten *prāṇas*.

Commentary

One speaks of four or ten *prāṇas* from the external, other-referential point of view (*vyavahāra naya*), but from the internal, self-referential point of view (*niścaya naya*), the Self or soul lives by the conscious vital force (*bhāva prāṇa*) or consciousness or the Self lives by the essence of life (*jīvatva*). The *bhāva prāṇa* (consciousness), which animates the physical vitalities and infuses life into the living organism, the body, mind and speech, senses and other physical vitalities. Without the *bhāva prāṇa*, the *dravya prāṇas* are unable to perform their functions, just as without electricity, electric gadgets, such as a fan, a bulb and an airconditioner are unable to perform their function.

31-32

अगुरुलहुगा अणंता तेहिं अणंतेहिं परिणदा सव्वे।
देसेहिं असंखादा सियलोगं सव्वमावण्णा॥

agurulahugā aṇaṁtā tehiṁ aṇtehiṁ pariṇadā savve ।
desehiṁ asaṅkhādā siyalogaṁ savvamāvaṇṇā ॥

केचित्तु अणावण्णा मिच्छादंसणकसायजोगजुदा ।
विजुदा य तेहिं बहुगा सिद्धा संसारिणो जीवा ॥

kecittu aṇāvaṇṇā micchādaṁsaṇakasāyajogajudā ।
vijudā ya tehiṁ bahugā siddhā saṁsāriṇo jīvā ॥

Jīvas (souls) are characterized by *agurulaghu* quality (see *PKS* 24) and have infinite spiritual qualities. They assume different forms through the manifestations of these infinite attributes. They have innumerable space-points. *Syāt*, i.e. in some respect, in a certain sense or from a particular point of view, the whole world is filled with their manifestations.

Some souls are affected by a deluded view, anger, passions, and the activities of mind, speech and body (*yoga*). Some others are free from these defects; these are numerous liberated souls (*siddhas*), while the other imperfect souls are mundane souls (*saṁsārī jīvas*).

Commentary

In the above verses, the author enumerates general and special characteristics of *jīva*, the characteristics both manifested and unmanifested.

33

जह पउमरायरयणं खित्तं खीरे पभासयदि खीरं ।
तह देही देहत्थो सदेहमत्तं पभासयदि ॥

jaha paumarāyarayaṇaṁ khittaṁ khīre pabhāsayadi khīraṁ ।
taha dehī dehattho sadehamattaṁ pabhāsayadi ॥

Just as the lotus-hued ruby, when placed in a cup of milk

imparts its lustre to the milk, so also the conscious Self (*ātmā*), residing in its own body, imparts its lustre of intuitive awareness and intelligence to the whole body.

Commentary

In this verse, the author explains the characteristic that the soul is of the same dimension as its own body, which is acquired through *karmas* by analogy.

34

सव्वत्थ अत्थि जीवो ण य एक्को एक्ककाय एक्कट्ठो ।
अज्झवसाणविसिट्ठो चिट्ठदि मलिणो रजमलेहिं ॥

savvattha atthi jīvo ṇa ya ekko ekkakāya ekkaṭṭho ।
ajjhavasāṇavisiṭṭho ciṭṭhadi maliṇo rajamalehiṁ ॥

The soul (*jīva*) pervades the whole body; or it exists in all the different bodies during different births; or that the souls are everywhere, i.e. in all parts of the world [in the last case, *jīvas* are not identical as they are distinct entities, yet they are one and identical (*ekkaṭṭho*) as they are substantially similar to one another, i.e. they are one from the generic point of view, but different from the specific point of view] (depending on the meaning of the word *sarvatra* in the verse). He is one with the body when functioning, yet he is not identical with it. Impelled by *adhyāvasāna bhāvas* (impure psychic dispositions), such as deluded views (*mithyātva*), lack of moral self-restraint (*anivrati*), perverted knowledge (*ajñāna*) and anger and passions, which are superimposed or transposed on the Self due to the coming into operation or rise (*udaya*) of material *karmas*, with the result that the Self is influenced or conditioned thereby, stained by the impurity of *karmas*, the soul puts on different forms in the transmigratory cycle of births and deaths.

Commentary

Though the soul is one with the body (*SS* 27), like milk and water, due to occupying the same space, they have quite distinct natures.

35

जेसिं जीवसहावो णत्थि अभावो य सव्वहा तस्स ।
ते होंति भिण्णदेहा सिद्धा वचिगोयरमदीदा ॥

jesiṁ jīvasahāvo ṇatthi abhāvo ya savvahā tassa ।
te hoṁti bhiṇṇadehā siddhā vacigoyaramadīdā ॥

The liberated souls (*siddhas*) do not have bio-energies or physical vitalities (*dravya prāṇas*) (including the five senses, the three energies of mind, speech and body, duration of life and respiration), yet they are not devoid of the vital conscious force (*bhāva prāṇa*), i.e. consciousness. In other words, the *jīva svabhāva* (intrinsic conscious nature) is not completely absent in the liberated souls. They are free from the corporeal limitations of the body; they are perfect souls beyond description.

Commentary by A. Chakravarti

The body is the limiting condition of *ātma-pradeśa* (dimension of *ātmā*). The size of the perfected soul is said to be a little less than that of its last body for it is free from the *yoga* – the principle which brings about contraction and expansion, or shrinking and diffusing. It may be said to be merely an assemblage of pure knowledge and other qualities of perfection.

36

ण कुदोचि वि उपण्णो जह्मा कज्जं ण तेण सो सिद्धो ।
उप्पादेदि ण किंचि वि कारणमवि तेण ण स होदि ॥

ṇa kudoci vi upaṇṇo jahmā kajjaṁ ṇa teṇa so siddho ।
uppadedi ṇa kiñci vi kāraṇamavi teṇa ṇa so hodi ॥

The liberated soul (*siddha*) is not to be born again at any time in *saṁsāra* (cycle of births and deaths). Hence, he is neither the effort (*kārya*) of *saṁsāra* nor he is the cause either, in the sense of bringing about a change in anything else.

Commentary

The *siddha* has secured the absolute, transcendental incorporeal (*amūrta*) state, the state of perfect spirituality and is incapable of maintaining causal relations with *saṁsāra* series. The latter is determined by kārmic conditions. The former, i.e. *siddha*, is not so conditioned.

37

सस्सदमध उच्छेदं भव्वमभव्वं च सुण्णमिदरं च ।
विण्णाणमविण्णाणं ण विजुज्जदि असदि सब्भावे ॥

sassadamadha ucchedaṁ bhavvamabhavvaṁ ca suṇṇa-midaraṁ ca ।
viṇṇāṇamaviṇṇāṇaṁ ṇa vijujjadi asadi sabbhāve ॥

The eight attributes of the soul are associated with the *siddha* state. These are (1) the soul substance as substance is permanent (*śāśvata*); (2) it is destructible every time with respect to modes (*paryāya*); (3) the substance is always capable of transformation or evolvent and as unprecedented consciousness and bliss, devoid of deterioration or change for the worse, (4) it will never evolve into delusion, attachment, impure psychic states, etc. (5) negation of extrinsic [other substance, other place, other time and other state or mode (*bhāva*)] qualities other than the innate pure state of the Self; (6) affirmation of its own innate pure nature of the Self, i.e. one's own (*sva*) substance, etc.; (7) that its knowledge is perfect, i.e. having omniscience; and (8) that it is devoid of or without the little or imperfect knowledge of sensory knowledge (*mati-jñāna*), etc.

Commentary by A. Chakravarti

As stated earlier, this verse defends the Jaina view of the Self against the nihilistic interpretation of the Buddhists. Even in the *siddha* state, the Self does not lose its *svabhāva* (substantial reality). It is only perfection of the already existing true nature which is in the germ in the finite Self. Complete evolution is not annihilation. There must be more of its real nature and not less.

With this ends the *amūrtta* characteristic of the *jīva*.

38

कम्माणं फलमेक्को एक्को कज्जं तु णाणमध एक्को।
चेदयदि जीवरासी चेदगभावेण तिहिवेण॥

kammāṇaṁ phalamekko ekko kajjaṁ tu ṇāṇamadha ekko ।
cedayadi jīvarāsī cedagabhāveṇa tihiveṇa ॥

One kind of *jīva* experiences only the fruits of *karma* (*karma-phala-cetanā*), such as pleasure and pain. Another kind of soul experiences the conative activity (*karma-cetanā*) as well; still others have knowledge consciousness (*jñāna-cetanā*). Thus, consciousness is manifested in threefold awareness.

Commentary by A. Chakravarti

The first two states of experience are related to the mundane soul (*saṁsārī jīva*), for they have reference to *karma*; whereas, the third has reference to pure consciousness nature (*cetanā svabhāva*) and hence is associated with the perfect souls. The implicit recognition by the author of the three different aspects of consciousness, feeling, activity and knowledge is worth noticing from the point of view of modern psychology.

39

सव्वे खलु कम्मफलं थावरकाया तसा हि कज्जजुदं।
पाणित्तमदिक्कंता णाणं विंदंति ते जीवा ॥

savve khalu kammaphalaṁ thāvarakāyā tasā hi kajjajudaṁ ।
pānittamadikkaṁtā ṇāṇaṁ viṁdaṁti te jīvā ॥

Indeed all the one-sensed, fixed living organisms (*sthāvara jīva*) like plants experience only feeling (*karma-phala-cetanā*); but the mobile, moving souls (*trasa jīva*) have, besides feeling, conative or activity experience (*karma cetanā*). Whereas those who transcend the organic conditions of bio-energies or physical vitalities (*dravya prāṇas*) experience pure knowledge consciousness (*jñāna-cetanā*).

Commentary by A. Chakravarti

Plants are fixed and incapable of movement, and can therefore only suffer environmental changes. They can only feel the mechanical and climatic stimuli around. The moving organisms because of the movement are capable of experiencing their own activity. In their experience then there is besides (1) *karma-phala-cetanā* (consciousness of the fruits of action), the affective state of experiencing the fruits of one's own past *karmas* or awareness of pleasure and pain; (2) that in which there is consciousness of activity (*karma-cetanā*); and (3) knowledge consciousness (*jñāna-cetanā*) in which there must be complete absence of physical and organic conditions. Such a being is certainly the *mukta jīva* (liberated soul).

After the wonderful discoveries of Jagdish Chandra Bose about plant life, it is not necessary to defend the proposition that plants are capable of feeling. Though Western science acknowledged the organic nature of plants, it was rather reluctant to admit the correlative organic characteristic of feeling in plants. This is only another phase of the Cartesian prejudice with reference to plants and animals. Descartes was responsible for the view that animals were skilfully created automata or machines. Charles Darwin gave a death blow to this philosophical superstition in regard to animals. Bose did the same service to the plant world and thereby distinctly established the fundamental unity of the organic world; and this is the thesis presupposed in the above verse.

40

उवओगो खलु दुविहो णाणेण य दंसणेण संजुत्तो ।
जीवस्स सव्वकालं अणण्णभूदं वियाणीहि ॥

uvaogo khalu duviho ṇāṇeṇa ya daṁsaṇeṇa saṁjutto ।
jīvassa savvakālaṁ aṇaṇṇabhūdaṁ viyāṇīhi ॥

Conscious attentiveness, psychic-attention, or manifestation or function of consciousness (*upayoga*) is twofold: *jñāna* (understanding or knowledge) and *darśana* (perception or intuition). They are always inalienable and present in *jīva*. So one should know.

Commentary by Amṛtacandra

Akalaṅka defines *upayoga* as "*caitanya anuvidhāyi pariṇāma* (states of evolution in obedience to consciousness)". It is of two kinds: *jñāna-upayoga* and *darśana-upayoga*. That which grasps particular (*viśeṣa*) is *jñāna* and that which grasps the general (*sāmānya*) is *darśana*; *upayoga* is always inseparable from *jīva* because they are made up of one existence (*astitva*).

Commentary by Jayasena

Upayoga is inseparable, integral or inalienable from *jīva* with respect to space-points (*pradeśa*), though they are different and distinct with respect to designation or appellation (*saṁjñā*), characteristics (*lakṣaṇa*), purpose (*prayojana*), etc.

41

आभिणिसुदोधिमणकेवलाणि णाणाणि पंचभेयाणि ।
कुमदिसुदविभंगाणि य तिण्णि वि णाणेहिं संजुत्ते ॥

ābhiṇisudodhimaṇakevalāṇi ṇāṇāṇi pañcabheyāṇi ।
kumadisudavibhaṅgāṇi ya tiṇṇī vi ṇāṇehiṁ saṁjutte ॥

Knowledge is of five kinds: (1) *abhinibodhake* or *mati–jñāna* (knowledge of the Self and the non-Self obtained by the senses and the mind); (2) *śruta* (scriptural knowledge); (3) *avadhi* (clairvoyance or visual, direct knowledge of matter); (4) *manaḥ-paryāya* (telepathic knowledge or mental knowledge, i.e. direct knowledge of another's mental activity); and (5) *kevala-jñāna* (perfect knowledge, omniscience, knowledge of all things and all their aspects at all times) (*TS* I.9). Of these five kinds of right knowledge, the first three, i.e. *mati* (sensory knowledge), *śruta* (scriptural) and *avadhi* (clairvoyance) can be erroneous knowledge characterized by doubt, perversity and carelessness, indifference or confusion (*TS* I.31), *kumati kuśruta* and wrong *avadhi* or *vibhaṅga.* Thus, there are eight kinds of knowledge.

[Additional Verses in Jayasena's Recension]

41.1 *Mati-jñāna* is of three kinds: (1) *uplabhdhi* (perception); (2) *bhāvanā* (contemplation); and (3) *upayoga* (psychic-attention). This *mati-jñāna* is also said to be of four kinds (*TS* I.15), viz. (1) *avagraha*, i.e. sense datum, perceiving or apprehension of a white object; (2) *ihā*, i.e. an attempt to determine what that white object is. This *ihā* refers to the indecisive mental attitude where several alternative determinations are possible; (3) *avaya*, i.e. judgement, the determining for sure that the white object is a pillar, not a man; and (4) *dhāraṇā* (retention). Thus, *jñāna* is always preceded by sense perception of presentation, i.e. *darśana*.

41.2 *Śruta-jñāna*, the objects of which may be physical or non-physical, is again of four kinds: (1) *labdhi*, i.e. attainment or the process of getting the meaning of one idea through its associated idea; (2) *bhāvanā*, i.e. contemplation; it is the direction of attention to one idea with a view to get at associated ideas; (3) *upayoga*, i.e. the process of understanding the meaning of ideas; and (4) *naya*, i.e. comprehension of a particular aspect of an object.

41.3 *Avadhi-jñāna* is of three kinds: *deśāvadhi*, *paramāvadhi* and *sarvāvadhi*. All the three are conditioned by psychic qualities. But *deśāvadhi* is also conditioned by birth in the case of a celestial being and the denizens of hell.

41.4 *Manaḥ-paryāya* is of two kinds: *ṛjumati* and *vipulamati*; telepathy which manifests straight and direct and telepathy that manifests crooked or undulations. These appear only in a person of *apramatta guṇasthāna* (stages of moral and spiritual development), who has acquired *saṁyama labdhi*, i.e. a person who has acquired complete harmony or steadiness of the spirit through renunciation.

41.5 *Kevala-jñāna* is knowledge par excellence. It is not conditioned by objects of knowledge. It is not even *śruta-jñāna* which is consequent upon study. To the person who has acquired perfect knowledge, there is no perversity of knowledge; it is only knowledge alone. So let it be known.

41.6 Due to the operation or rise (*udaya*) of delusion (*mithyātva*), knowledge gets vitiated and becomes perverse knowledge (*ajñāna*) and also lacks moral self-restraints (*avirati bhāvas*). Thus, it obscures an enlightened world-view or right belief; and due to *mithyātva*, depending on *jīva*, etc. knowable objects, at the time of investigating the reality of principles (*tattvas*), the viewpoints and valid knowledge (*pramāṇa*) become misleading and untrustworthy.

Commentary

Mithyātva (delusion or deluded view of the reality of things) is the condition of *ajñāna* (wrong or perverse knowledge). Its presence in a soul is responsible for several undesirable consequences. This *mithyātva* may be innate or acquired. It accounts for the loss of the enlightened world-view, right faith and loss of right or enlightened knowledge.

The deluded view is of five kinds:

1. one-sided (*ekānta*), i.e. taking only one aspect of a many-sided thing, e.g. man is mortal or immortal, regarded from the point of view of body and soul respectively;
2. perverse belief (*viparīta*), e.g. animal sacrifices lead to heaven;
3. doubt (*saṁśaya*), i.e. scepticism or hesitation, e.g. as to the path of liberation;
4. veneration (*vinaya*), i.e. taking all religions and gods to be equally worthy of pursuit; and
5. perverse knowledge or nescience (*ajñāna*). Wrong belief caused by ignorance and indiscrimination between good and bad. It is caused by inborn error (*nisarga*) or by preaching of other (*adhigama*).

The last (*ajñāna*) is of four kinds:

a. *kriyāvādī*, i.e. belief in time, soul, etc. as causing everything in the world;
b. *akriyāvadī*, opposite of that;
c. *ajñānika*, i.e. agonistic; and
d. *vaināyika*, i.e. religion is identical with veneration of parents, king, etc., e.g. Confucianism.

All these four together form 363 varieties of wrong belief (or *pākhaṇḍa*), e.g. 180, 84, 67 and 32 respectively.

Nisargaja is also called *agrahīta* and *adhigamaja* is called *grahīta*.[8]

---

[8] J.L. Jaini, *Tattvārthasūtra*, p. 130 (on *TS* VIII.1).

42

मदिणाणं पुण तिविहं उबलद्धी भावणं च उवओगो।
तह चेव चदुवियप्पं दंसणपुव्वं हवादि णाणं॥

madiṇāṇaṁ puṇa tivihaṁ ubaladdhi bhāvaṇaṁ ca uvaogo ।
taha ceva caduviyappaṁ daṁsaṇapuvvaṁ havādi ṇāṇaṁ ॥

Darśana (perception or intuition) is of four kinds: (1) perception through visual sensations (*cakṣu-darśana*); (2) perception through non-visual senses (*acakṣu-darśana*); (3) through the faculty of *avadhi* (clairvoyance); and (4) through *kevala* (infinite perception), which is unlimited and apprehends all reality. Thus, it is described.

Commentary by A. Chakravarti

This *mati-jñāna* (knowledge of the Self and the non-Self obtained by the senses and the mind) is achieved only upon sense perception. It includes comprehension of the object, memory, and understanding, i.e. all that is given to us through sense perception and all that we elaborate out of these sense elements in memory and imagination. Though it is mainly of three kinds, it also has four different forms. For example, the author of the *Tattvārthasūtra* speaks of *avagraha* (sense datum) *dharna* (retention). This fourfold division is not fundamentally different from the previous division (*avagraha*), which implies the sense datum. For example, perceiving a thing to be white through the eyes is *avagraha*. To attempt to determine what that white object is, is *ihā*. This *ihā* refers to the indecisive mental attitude where several alternate determinations are possible. Finally, when the thing is determined or chosen out of several alternatives because of certain special characteristics perceived in the white object, then we have *avaya* (judgement). This implies the inferential element in all perception. This should not be confounded with mediate inference about other things through sense perception. The very same act of perception involves all these three stages. And lastly, whenever we remember these things after sometime, it is called *dhāraṇā*. This *mati-jñāna* consists of 336 forms when viewed according to different principles of *avagraha*.

43

सुदणाणं पुण णाणी भणंति लद्धी य भावणा चेव ।
उवओगणयवियप्पं णाणेण य वत्थु अत्थस्स ॥

sudaṇāṇaṁ puṇa ṇāṇi bhaṇanti laddhī ya bhāvaṇā ceva ।
uvaogaṇayaviyappaṁ ṇāṇeṇa ya vatthu atthassa ॥

The soul is not distinct from its attribute of knowledge and because the ways of knowledge are many, the world of reality or the universe is also said to be multi-verse by the wise.

Commentary

The verse emphasizes the inalienability of the soul or the Self from its attributes.

44

ओहिं तहेव घेप्पदु देसं परमं च ओहिसव्वं च ।
तिण्णिवि गुणेण णियमा भवेण देसं तहा णियदं ॥

ohiṁ taheva gheppadu desaṁ paramaṁ ca ohisavvaṁ ca ।
tiṇṇivi guṇeṇa ṇiyamā bhaveṇa desaṁ tahā ṇiyadaṁ ॥

If the substance (*dravya*) is entirely (*sarvathā*) separate, different or distinct (*vibhakta*) from its attributes or qualities (*guṇas*), then it may be transformed into infinite other substances or if the *guṇa* can exist separate or distinct from its own substance, there will be no necessity for a substance at all. (In either way, the result would be absurd.)

Commentary by A. Chakravarti

The relation of substance and its qualities is an extremely difficult problem for a metaphysician. Some would emphasize the qualities to the detriment of substance and some would emphasize the substance at the cost of qualities. In either case, the result would be absurd because of the false emphasis.

We may have the world of unchanging reality of the Greek philosopher Parmenides or a world of eternal flux of Heraclitus, another Greek philosopher, but both these worlds are so far removed from the world of our concrete experience. In our world of concrete experience, things and qualities are of equal importance. In fact, the distinction is not absolute, it is relative.

There can be no qualities (*guṇas*) apart from substance nor substance (*dravya*) without qualities. If substance is deprived of all its qualities and if it is still the possible substratum of qualities, then different groups of qualities may get associated with that substratum, i.e. the same substance has the chance of becoming infinite other substances. This is manifest absurdity. Or again, if the attributes can float themselves without any fixing substratum then they would by their own combination constitute a thing and the substance is no more necessary. To think of reality as identical (*ekatva, ananya*) with qualities apart from the qualified thing is also equally absurd. On account of internal contradiction the view that qualities may be different or distinct (*vibhakta*) from substance is criticized. The arugment adopted by the author reminds one of George Berkeley and David Hume.

Berkeley's argument against John Locke's doctrine of substance similarly emphasizes the absurdity of an unknown *x* which is substance for Locke. This unknown *x* may get several determinations and thus may become several distinct things. But Berkeley's main argument is slightly different from the author's. His attack on substance is based upon the absurdity of abstract ideas. Again, Hume after reducing the world to a group of sense-presented ideas points out the absurdity referred to in the latter half of the *gāthā* (verse) by his doctrine that anything may be the cause of anything else.

Similar tendencies are not altogether absent in Indian thought. You have the one-sided emphasis both in Vedāntism and Buddhism. The idea of *nirguṇa* reality and the principle of Kṣaṇikavāda are the two rival doctrines; and both are rejected by Kundakunda; qualityless reality and the qualities bringing about a new reality every moment are both untenable according to the author.[9]

45

विउलमदी पुण णाणं अज्जवणाणं च दुविह मणणाणं।
एदे सजमलद्धी उवओगे अप्पमत्तस्स॥

viulamadī puṇa ṇāṇaṁ ajjavaṇāṇaṁ ca duviha maṇaṇāṇaṁ ।
ede sajamaladdhi uvaoge appamattassa ॥

Those who know the truth do not recognize any of the following views as to the relation of substance and quality: (1) that they are (absolutely) identical (*ananya*) or non-distinct (*avibhakta*); (2) that they are [absolutely] different or distinct (*vibhakta*) and other (*anya*) in every way; and (3) that they are both identical and distinct at the same time.

Commentary

Here in this verse Kundakunda wants to make it clear that substance and quality are not distinct (*avibhakta*) and other (*anya*) in *rerum-natura* (the nature of things); but only in thought and, there too, the distinction is only relative, i.e. the quality cannot be thought of as absolutely (*sarvathā*) independent of substance. Still, it can be considered to be as distinct (*vibhakta*) from the thing.

46

णाणं णेयिणिमित्तं केवलणाणं ण होदि सुदणाणं।
णेयं केवलणाणं णाणाणाणं च णत्थि केवलिणो॥

9 A. Chakravarti, 1971, *Samayasara of Kundakunda*, Introduction, pp. 41-42, Delhi: Bhartiya Jnanapeeth.

ṇāṇaṁ ṇeyiṇimittaṁ kevalaṇāṇaṁ ṇa hodi sudaṇāṇaṁ ।
ṇeyaṁ kevalaṇāṇaṁ ṇāṇāṇāṇaṁ ca ṇatthi kevaliṇo ॥

The difference or distinctness (*vibhaktapana*) between *dravya* and *guṇa* is determined by nomenclature, appellation or name (*vyapadeśa*), form, appearance (*ākāra*) or expression (*saṁsthāna*), number (*saṅkhyā*) and locale, locality or subject (*viṣaya*). These various determinants hold good among things (*dravya*) and their attributes or qualities (*guṇas*) whether these are considered identical (*ananya*) [i.e. having *ekatva*] or different (*anya*) [i.e. distinct (*vibhakta*)]. These differences hold good not only among different things but also in the case of a single complex thing which is capable of internal distinction.

Commentary by A. Chakravarti

The difference between substance and qualities due to *vyapadeśa*, etc. of this verse can be explained as follows.

Vyapadeśa difference, i.e. difference due to nomenclature, appellation or name is of two kinds: (1) *ṣaṣṭhī vyapadeśa*, i.e. substratum or possessive relation (*adhikaraṇa*) difference; and (2) difference of relation of causal factors (*kāraka vyapadeśa* or *kāraka saṁjñā;* there are six *kārakas*).

To illustrate the different aspects of the causal relation (the six *kārakas*) that exist among different things, it may be said: *devadattaḥ phalam aṅkuśena dhanadattaya vṛkṣāt vāṭikāyām avacinoti*, i.e. "Devadatta, the agent or subject (*kartā*) plucks a fruit (*karma* the object or effect) for Dhanadatta (*nimitta*, purpose, *sampradāna*) from a tree (*sakāśa*, the place from which the effect issues, *apadāna*) in his orchard (substrate, the place in which the cause operates, *adhikaraṇa*) with a hook (the instrument, *kāraṇa*)." This relates to several independent things.

The same causal relation (*kāraka-saṁjñā*) having six elements existing in the same identical complex thing may be illustrated as: *tadeva ātmā (kartā) ātmānam karmatāpanna ātmanā (kāraṇabhūtena) ātmane nimittam ātmanaḥ sakāśāt ātmani adhikaraṇabhūte dhyāyati* (*PKS* 46 JS), i.e. "The Self, i.e. *ātmā* (*kartā*) about himself (*karma*) with his self-reflection (*kāraṇa*) for his own purpose (*nimitta*, *sampradāna*) drawing out of himself (*sakāśa*, *apādāna*) and yet reposing in himself (*adhikaraṇa*) contemplates." Here the causal relation is with reference to the same complex thing.

Each of these holds good in case of different things as also in the case of a single complex thing.

1. *Vyapadeśa*: The difference between the Vindhyācala and the Himalayas is the difference of nomenclature or name. The possessive relation among different things is illustrated as "Mohan's cow". The same relation is illustrated by a single complex thing as "The branch of a tree" or a Jina's attributes. Here the possessive relation is distinctly internal in the very same thing.
2. *Saṁsthāna* (*ākāra* or expression): This determinant is applied to different things as in Mohan's tall cow. It is also applied to the same thing as illustrated in "The tall branch of a tall tree".
3. *Saṅkhyā* (number): The statement "Mohan's ten cows" signifies a quantitative difference between two distinct things – Mohan and cows. But the quantitative difference may exist internally in the same thing as the ten branches of the tree or the many attributes of *dravya* (substance).
4. *Viṣaya* (basis, substratum or locative difference). "In the cowshed is the cow" signifies that the locality (i.e. the place in which the cause operates) is external. The internal locative relation is expressed in the statement "In the substance are the qualities".

Thus, the difference due to *vyapadeśa*, *saṁsthāna*, etc. is seen among different things as well as in the same thing, which is internally differentiated (*anyatva*). Hence, when one kind of relation is implied it need not be confounded with the other. If substance and attributes or qualities are said to be different, this difference is not to be interpreted that distinct (*vibhakta*) and other (*anya*). Hence, the relation between substance and attributes can certainly be from one aspect a relation of difference [i.e. from the external, other-referential point of view (*vyavahāra naya*)] and still the two need not be absolutely distinct. In short, the relation between substance and attribute is one of identity and difference. That this relation of identity in difference is not self-contradictory is supported by analogy.[10]

Pañcāstikāyasāra 46 explains that when a thing is described with reference to only one substance and its attributes or qualities, then *vyapadeśa* (appellation), etc. four factors or aspects are said to be spoken of as one of identity or unity (*ekatva*). But when description pertains to many (*aneka*) or different substances, the description with reference to *vyapadeśa* (appellation), etc. four factors or aspects are regarded as distinct (*vibhakta*).

1. *Vyapadeśa*: To illustrate the verbal description (*vyapadeśa*) in respect of identity, we may say "The branch of a tree" because the branch is not different or distinct from the tree. But to say "Mohan's cow" is a verbal description (*vyapadeśa*), which refers to distinctness (*vibhakta*) as a cow is different from a man, viz. Mohan.

2. *Saṁsthāna* (shape or size, *ākāra*): To say "The big branch of a big tree" is identical, i.e. *ekatva saṁsthāna*, but to speak of "Mohan's big cow" is indicative of distinct (*vibhakta*) *saṁsthāna*.

3. *Saṅkhyā* (number): "Ten branches of the tree" is an identical (*ekatva*) description of number, but "Mohan's ten cows" is the description of distinctness (*vibhakta*) in number as Mohan and the cow are two different entities.

---

10 A. Chakravarti, 1971, *Samayasara of Kundakunda*, Introduction, pp. 43-44, Delhi: Bhartiya Jnanapeeth.

4. *Viṣaya* (locative difference): "The branches in a tree" is identical (*ekatva*) or non-distinct (*avibhakta*) description of substance (*dravya*) and qualities (*guṇas*) *viṣaya.*

The above-mentioned four factors are said to be the nature of identity (*ekatva*) in *dravya* and *guṇa* and they are descriptions of distinctness (*vibhakta*) in different or other (*anya*) substances.

The statement of Naiyāyika that belief in identity (*ekatva*) of *dravya* and *guṇa, vyapadeśa, saṁsthāna,* etc. is not proved, is not correct.[11] There can be descriptions of *vyapadeśa,* etc. in both distinctness and identity.

Besides *Pañcāstikāyasāra* 45 and 46 discussed above, *Pañcāstikāyasāra* 43-44 and 47-52 are also relevant in regard to the concept of *ekatva-vibhakta* with reference to *dravya, guṇa* and *siddha.* The soul substance (*jīva dravya*) is not distinct from its attribute (*guṇa*) like knowledge (*PKS* 43). There can be no knowledge apart from the knower. The Self and knowledge are not distinct (*avibhakta*). In *Pañcāstikāyasāra* 13, it is clearly stated that in the world there are no qualities (*guṇas*) without substance (*dravya*) and *guṇas* are not distinct (*vibhakta*) from *dravya* as *dravya* is the substratum of *guṇa.*

The relationship of ownership and possession may exist between two different things or in the same thing internally differentiated. Hence by analogy Kundakunda brings out the *ekatva–vibhakta* aspect of the relation.[12]

47

मिच्छत्ता अण्णाणं अविरदिभावो य भावआवरणा।
णेयं पडुच्च काले तह दुण्णय दुप्पमाणं च ॥

[11] See Jagdish Prasad Jain (ed.), 2005, *The Enlightened Vision of the Self: Svarupa Sambodhana* of *Akalanka*, tr. D.K. Goyal, Foreword, p. 5, New Delhi: Radiant Publishers.

[12] Introductory remark of Amṛtacandra to *Pañcāstikāyasāra* 47.

micchattā aṇṇāṇaṁ aviradibhāvo ya bhāva-āvaraṇā ।
ṇeyaṁ paḍucca kāle taha duṇṇaya dupyamāṇaṁ ca ॥

Just as *dhana* (wealth) and *jñāna* (knowledge) make the owners *dhanī* (rich, possessor of wealth) and *jñānī* (possessor of knowledge), thus expressing two ways of relationship (unity (*aiyatta* in Prākr̥t and *ekatva* in Sanskrit) and distinctness (*vibhakta*), so also the relation between substance and its qualities implies both the aspects of identity (*ekatva*) and difference or distinctness (*vibhakta*). Thus, say those that know the truth.

Commentary by A. Chakravarti

Where there are different existent realities (*astitva nivr̥ttam*) or different substances having different forms (*saṁsthāna*), different numerical properties (*saṅkhyā*) and different subject, locality or basis (*viṣaya*) as in the case of wealth (*dhana*) and the person owning wealth (*dhanī*), the relation is one of separateness (*prathakatva* or distinctness) as the two (*dhana* and *dhanī*) have their different existence or verbal descriptions (*vyapadeśa*), *saṁsthāna*, *saṅkhyā* and *viṣaya*. But the relation between knowledge (*jñāna*) and person having knowledge (*jñānī*) is one of identity or unity (*ekatva*).

One who owns *dhana* (wealth) becomes on account of that ownership a *dhanī* (rich). Here too, different things because of the relation of possession are brought together, i.e. because of possession there is union and again because of possession the things uniting must be different. Things may exist as separate and distinct when the relation may unite them or there may be single things which because of the relation may get differentiated. The illustration of *dhana*, *dhanī* is of the former kind. The illustration for the latter is *jñāna*, *jñānī*. The person to start with is one, but on account of this relation of possession the single entity gets differentiated. In order to be called *jñānī*, one who possesses *jñāna* (knowledge) the possessed thing must be differentiated from the owner. Otherwise, the relation of ownership will have no meaning. Hence, we have to admit that the relation of ownership implies both unity (*ekatva*) and diversity [or distinctness (*vibhakta*)] whether the things related are different or identical in themselves. Thus, the author concludes that the relation between substance and its qualities exhibits both these characteristics. It is not coherent to associate both the characteristics with the relation.

Commentary by Jayasena

As compared to wealthy and rich persons, which have different existent reality, knowledge and the knower have same or identical existences, their names, forms, number and substratum are identical. Thus, they are inseparable.

48

दंदणमवि चक्खुजुदं अचक्खुजुदमवि य ओहिणा सहियं।
अणिधणमणंतविसयं केवलियं चावि पण्णत्तं॥

daṅdaṇamavi cakkhujudaṁ acakkhujudamavi ya ohiṇā sahiyaṁ ।
aṇidhaṇamaṇaṅtavisayaṁ kevaliyaṁ cāvi paṇṇattaṁ ॥

If the Self (knower) and its knowledge (*jñāna*) are always substantially distinct one from the other, then each will become a non-conscious entity. That view being self-contradictory would undoubtedly conflict with the truth revealed by the Omniscient Lord (Jina).

Commentary by Amṛtacandra

The attributes are always dependent on substance (*TS* V.41) (which acts as the substratum of its attributes) and cannot exist without the substance, i.e. the attributes (*guṇas*) are never without *dravya*. It follows that if that *dravya* is different (*anya*) from *guṇa*, then the attributes (*guṇas*) will still be dependent on some (*dravya*). The attributes are infinite. And if that substance is also different (*anya*) from the attributes (*guṇa*), then in that case due to the difference between substance and attributes, there will be infinite substances (*ananta dravya*) (*PKS* 44 AC). Thus, when *guṇa* and *dravya* are separated not only each becomes a meaningless abstraction incapable of existence, but also there will be new *dravyas* with the result that there will be infinite substances.

Moreover, in reality the substance (*dravya*) is the collection (*samudaya*) of attributes (*guṇas*). If the attributes (*guṇas*) are different (*anya*), then substance will be no more, will not be existent, i.e. there will be no necessity for substance at all (*dravya abhāva*) (*PKS* 44 AC).

Besides the above-mentioned two faults in case of the Self and its knowledge or substance and attributes being substantially, absolutely or entirely distinct (*vibhakta*) or different (*anya*) one from the other, then each will become non-conscious (*acetana*) or a non-spiritual entity, due to knowledge being without its agent (*kartā*) part (*aṁśa*) (A. Chakravarti).

That view being self-contradictory would undoubtedly conflict with the truth revealed by Jina. Here, it may be clarified that *jinavamatam* (view of Jina) (in *PKS* 48) is not to be taken as the reason for rejecting the opposite view. The real standard of truth (*pramāṇa*) is not the principle of revelation. Though the Jaina system of thought is very often and undoubtedly referred to as revealed by Jina, the Jaina system is acknowledged to be true not because it is revealed by a great spiritual being (Jina), but because the revelation is borne out by the nature of Reality. Jaina thinkers therefore attach more importance to the absence of self-contradiction than to revelation (A. Chakravarti).

Pañcāstikāyasāra 49-50 pertain to the refutation of the Mīmāṁsā and Vaiśeṣika systems of philosophy that knowledge is a distinct entity from the soul and that the soul becomes the knower due to combination (*samavāya*) with the body, mind and senses.[13]

49

ण वियप्पदि णाणादो णाणी णाणाणि होंति णेगाणि ।
तम्हादु विस्सरूवं भणियं दवियत्ति णाणीहि ॥

ṇa viyappadi ṇāṇādo ṇāṇī ṇāṇāṇi hoṁti ṇegāṇi ।
tamhādu vissarūvaṁ bhaṇiyaṁ daviyatti ṇāṇīhi ॥

If the Self is entirely distinct (*vibhakta*) in nature from knowledge, then he cannot become, even by combination (*samavāya*) with knowledge or thought, *jñānī* (knower) or the thinker. "Unthinking thing" will still be the same denoting the identity or unity (*ekatva*) brought out by the combination of substance and quality which are in themselves unthinking things.

Commentary

Thus, if the Self and knowledge be really distinct, then the two would never be able to constitute a single spiritual being even by the process of combination.

Surely, the Self, having the quality of knowledge (*jñāna guṇa*) through his *jñāna guṇa*, even when knowing the entirely different other (*para*) things, does not become knower (*jñānī*) by the relation of combination (*samavāya*) of other (*para*) things. Even the statement that the *jīva* (Self) is *ajñānī* (ignorant of knowledge) proves the identity or unity (*ekatva*) of *guṇa* (quality) and *guṇī* (possessor of *guṇa* (quality)). The *guṇī dravya* (knowing substance) is never different or distinct (*vibhakta*) from its *guṇa* which is intrinsic to its nature and simultaneous (*svabhāvī, sahavartī*). This verse refutes the view of Vaiśeṣikas who believe that the Self (*jīvātmā*) was different

13 Introductory remarks of Amṛtacandra to *Pañcāstikāyasāra* 49-50.

from *jñāna guṇa* (quality of knowledge) and later became *jñānī* due to the relation of combination of related things.

Akalaṅka disagrees with the Mīmāṁsakas and the Vaiśeṣikas, who maintain that *jñāna* (knowing capacity) is not the nature of the soul; it is an accidential or adventitious quality generated in the soul on account of its contact with the body, mind and sense organs. *Jñāna* is, thus, considered to be a distinct entity from the soul and is said to be brought in association with the soul (*jīvātmā*) by combination (*samavāya*); then the soul becomes the knower. In other words, the quality of consciousness (*cetanā*) and the substance of the soul (*ātma-dravya*) are two different entities occasionally or accidentally brought together by extraneous circumstances.

Commentary by A. Chakravarti

If the Self and knowledge or thought are in themselves unthinking things, then by combination they would still be unthinking. Consciousness is not to be derived from non-conscious elements existing severally or in combination. If thought or knowledge is an adventitious characteristic of the Self, then before the acquisition of knowledge or thought he must have been either a *jñānī* or *ajñānī* (thinking or unthinking). If one is thinking then the combination with *jñāna* (knowledge) is unnecessary and useless. But if one is unthinking then, is that attribute "unthinking" again innate or adventitious? If it is adventitious, then the Self cannot acquire thought or knowledge for this attribute is incompatible with his former acquisition. If the unthinking quality is innate, then the thinking quality may as well be taken to be innate. Thought or knowledge then is not an adventitious attribute of the Self. The Self is *jñānī* not as a result of combination of the Self and knowledge or thought. Even the perfect knowledge omniscience (*kevala-jñāna*) is innate in the soul, though it is veiled for the time being by the *karmas*.

50

जदि हवदि दव्वमण्णं गुणदो य गुणा य दव्वदो अण्णे।
दव्वाणंतियमधवा दव्वाभावं पकुव्वंति॥

jadi havadi davvamaṇṇaṁ guṇado ya guṇā ya davvado aṇṇe ।
davvāṇaṅtiyamadhavā davvābhāvaṁ pakuvvaṅti ॥

The relation between substance and its quality is one of coeval (*samavartī*), identity and unity (*ekatva*) or inseparability (*apr̥thakbhūta*). That is why the unity (*ekatva*) of a substance and its qualities is said to be not the result of the union or a combination (*samavāya*) of two independent elements. Therefore, it has *ayuta-siddhattvam* not separateness (not *pr̥thagbhūtatva*).

Commentary by Amr̥tacandra

Since the relation between substance and quality is that of identity or unity (*ekatva*), the explanation between them by the principle of combination (*samanvaya*) is thus shown to be unwarranted and absurd.

Commenting on this verse Amr̥tacandra observes:

> As *dravya* (substance) and qualities (*guṇa*) are made, formed or created (*nivr̥ttitvāt* or *racita*) from one and same existence (*astitva*), therefore, their beginningless (*anādi*) infinite (*ananta*) simultaneity (*sahavartī*) is in reality co-eval (*samavartī*); the same is a combination (*samavāya*) in Jaina view. In spite of name (*saṁjñā*), etc. [differences] because of identity with reference to belonging to the same identical complex thing (*vastutvena abhedāt*), there is inseparability (*apr̥thagbhūtatvam*). Then due to the absence (*abhāva*) of different existence, which is the cause of *yutsiddhi*, there is *ayutsiddhatvam*. Therefore, substance and qualities being the combination (*samavāya*) having the

simultaneity (*samavartitva*) it is only *ayutsiddhi*, not separateness (*pṛthagbhūtatvam*) (*PKS* 50 AC).

Commentary by A. Chakravarti

The term *samavāya* (union or combination) implies mainly unity and also union. Union implies the existence of independent elements which constitute the unity. The elements must be prior to the unity. Then unity will be the result of combination. The relation between *guṇa* (quality) and a person possessing quality (*guṇī*) is not one of combination.

Hence though the author designates the relation by the term *samavāya*, he strictly excludes the other implications of the term. The term means nothing more than oneness or unity (*ekatva*). *Dravya* and *guṇa* (substance and quality) are merely the different aspects of the same reality and as such they must exist together without beginning and without end.

This eternal coexistence is implied by the term *samavāya*. This coexistence of *dravya* and *guṇa* is termed *samavāya* (unity) by Kundakunda. The very same relation implies non-distinctness or inseparability of the two though they are different in name. Hence, the relation is called *apṛthagbhūtatvam*. Again since the unity is not brought about by the combination of two independent elements, it has *ayut-siddhatvam*. Therefore, the relation is one of unity (*ekatva*) and not of union (*samavāya*) as the Mīmāṁsakas and Vaiśeṣikas believe.

51-52

अविभत्तमणण्णत्तं दव्वगुणाणं विभत्तमणण्णत्तं ।
णिच्छंति णिच्चयह्णू तव्विवरीदं हि व तेसिं ॥

avibhattamaṇaṇṇattaṁ davvaguṇāṇaṁ vibhattama-ṇṇattam ।
ṇicchaṁti ṇiccayahṇūṁ tavvivarīdaṁ hi va tesiṁ ॥

वववदेसा संठाणा संखा विसया य होंति ते बहुगा ।
ते तेसिमणण्णत्ते अण्णत्ते चावि विज्झंते ॥

vavadesā saṁṭhāṇā saṅkhā visayā ya hoṅti te bahugā ।
te tesimaṇaṇṇatte aṇṇatte cāvi vijjhaṅte ॥

Pañcāstikāyasāra 51-52 state: Colour, taste, smell and touch are the qualities of the primary [or the subtlest] atom (*paramāṇu*). They are not said to be really distinct from their material substratum though they are undoubtedly distinguishable from it as regards name, *saṁjñā*, etc. (*PKS* 51-52 AC), verbal description (*vyapadeśa*), form (*saṁsthāna*), etc. (*PKS* 47) [they are not other (*anya*) in nature (*svabhāva*)]. Similarly, perception (*darśana*) and understanding [or knowledge] (*jñāna*) are really identical or not other (*ananya*) from the Self and are not distinct from it; though from the point of view of [name, *saṁjñā*, etc.] (*PKS* 51-52 AC), verbal description (*vyapadeśa*), form (*saṁsthāna*), etc. they may be spoken of as distinct (*vibhakta*) from the substratum Self, but not distinct in nature. In short, though distinguishable in thought the faculties of *darśana* and *jñāna* are not really distinct (*vibhakta*) from the Self.

Commentary by A. Chakravarti

The author [Kundakunda] establishes the proposition that distinction in thought does not necessarily imply distinctness in nature. What may be distinguished by comparison may in reality exist in essential unity. This result is obtained by the principle of analogy. In the case of matter, the qualities are not distinct and separate from the substratum, though they may be referred to by different names, different forms and so on. Exactly in the same way are the conscious qualities of perception and understanding related to *jīva*. As attributes they are distinct from the ego [Self] and may be distinguished by name, form, number, etc. But, still they are not really distinct from it.

Distinctness necessarily presupposes an underlying identity between the things compared. Apart from this identity there can be no comparison and distinction. What are distinct must really be identical. Hence, distinction in thought, instead of presupposing distinctness in nature, implies identity. This view of Kundakunda is borne out by modern thought.[14]

53

णाणं धणं च कुव्वदि धणिणं जह णाणिणं च दुविधेहिं।
भण्णंति तह पुधत्तं एयत्तं चावि तच्चण्हू ॥

ṇāṇaṁ dhaṇaṁ ca kuvvadi dhaṇiṇaṁ jaha ṇāṇiṅaṁ ca duvidhehiṁ ।
bhaṇṇaṅti taha pudhattaṁ eyattaṁ cāvi taccaṇhū ॥

The soul has five primary or fundamental qualities; they are existences infinite in number. [Of these, five qualities (called the soul's own spiritual principles (*sva-tattvas*) in *TS* II.1) or psychic dispositions (*jīva bhāva*) the innate pure natural *bhava*, viz. (1) *pariṇāmika bhāvas* have neither beginning nor end; [the three (2, 3 and 4), viz. (2) subsidential (*aupaśamika*), (3) partly destructive and partly subsidential (*kṣayopaśamika*), and (4) the coming into operation or rise (*udaya*) of *karmas*, i.e. *audāyika*], have beginning and end; [and the fifth, i.e. destructive (*kṣāyika bhāva*), has a beginning, but no end.

Commentary by A. Chakravarti

Jīvas really are the agents, since they bring about different modes with different names. They are described as (1) without beginning or end, (2) with a beginning and end, and (3) with a beginning and no end.

---

[14] A. Chakravartinayanar, 1975, *Panchatiskayasara: The Building of the Cosmos*, p. 48, New Delhi: Bhartiya Jnanapeeth.

1. If we attend to the essential, innate pure nature of *jīvas* (*pariṇāmika bhāva*), i.e. their thought or consciousness, then they may be said to have neither beginning nor end; as spiritual existence they are eternal. Neither are they created nor can they be destroyed.
2. The very same *jīvas* from the psychic qualities of *audāyika*, i.e. coming into operation, rise (*udaya*) or fruition of *karmas*; *aupaśamika* (subsidential) and *kṣayopaśamika* (partly destructive and partly subsidential) have both beginning and end.
3. But from the *kṣāyika bhāva* (destructive state), they have beginning but no end. It cannot be said in the last case that because there is beginning there must also be an end. When they are free from *upādhis*, then they realize their true nature and they become *siddhas* (liberated souls). *Jīvas* that are found with such characteristics are infinite in number. Again, it cannot be said that in the case of these *jīvas* which really have neither beginning nor end, the other characteristics would be inconsistent. The other characteristics are having both beginning and end, and having beginning but no end. These characteristics are true of *jīvas* because of kārmic entanglements. The presence of *karma* with *jīva* is the condition of *sāṁsārika* (mundane) transformations and *jīvas* in *sāṁsārika* cycles have both birth and death. But in the last stage towards emancipation from *karmas*, *jīvas* may be said to have only birth but no death. For, there is spiritual immortality as a result of complete emancipation.

The five fundamental characteristics implied in this verse are the five *jīva bhāvas*. These are: (1) *audāyika*, (2) *aupaśamika*, (3) *kṣayopaśamika*, (4) *kṣāyika*, and (5) *pariṇāmika*.

The last one is the attribute of consciousness. This conscious nature is eternal. Hence, *jīva* is *anādi-nidhana* (without beginning or end). The first three bring about *sāṁsārika* changes of birth and death. Therefore, from their aspect, *jīva* has both beginning and end. The fourth is the characteristic of emancipation. The emancipated Self has no beginning and no end.

54

णाणी णाणं च सदा अत्थंतरिदा दु अण्णमण्णस्स ।
दोण्ह अचेदणत्तं पसजदि सम्मं जिणावमदं ॥

ṇāṇī ṇāṇaṁ ca sadā atthaṅtaridā du aṇṇamaṇṇassa ।
doṇha acedaṇattaṁ pasajadi sammaṁ jiṇāvamadaṁ ॥

Thus, *jīvas* that are, may die and those that are not, may be born; thus says the Omniscient Lord (Jina). Though the statements (this verse and the nineteenth) are apparently conflicting, they are not contradictory.

Commentary by A. Chakravarti

From *dravyārthika naya*, i.e. from the substance aspect of essential nature, *jīvas* can have neither birth nor death. But from *paryāyārthika naya*, from the point of view of modifications, they have births and deaths. The two propositions are quite compatible with one another and they state different aspects of *jīva*. The other proposition with which this verse appears to be in conflict is verse 19 which states: "There is no death for the existing *jīvas*, nor birth of the non-existing ones." This certainly appears to contradict the present *sūtra* (verse) and the commentator establishes that the conflict is only apparent and not real.

55

ण हि सो समवायादो अत्थंदरिदो दुणाणादो णाणी ।
अण्णाणीति च वयणं एगत्तप्पसाधगं होदि ॥

na hi so samavāyādo atthaṅdarido duṇāṇādo ṇāṇī ।
aṇṇāṇīti ca vayaṇaṁ egattappasādhagaṁ hodi ॥

Life is hell, life has plant and animal, as a man or as a celestial being (*deva*), the states of being are caused by the respective *nāma-karma prakṛtis*. These bring about death to *jīvas* that are and birth to those that are not.

Commentary by A. Chakravarti

The different states of existence are the different modifications of *jīva* brought out by *upādhi*, i.e. the limiting kārmic conditions. Those *karmas* that determine the next individual state of *jīvas* are *nāma-karmas*. These *nāma-karmas* are the real causal conditions that lead the soul to manifest in a particular form. These forms appear and disappear. Birth and death are characteristics associated with these forms or modes. Just as waves may appear and disappear on the surface of the ocean as a result of the wind, while the base or the bottom of water is really unaffected, so *jīva* remaining eternal and constant in its real essence is the basis of the different waves of existence appearing and disappearing. And these waves in the ocean of life are brought about by *upādhis* or kārmic conditions. The characteristics of birth and death which are really true of the form of existence are also predicated of *jīva*.

56

समवत्ती सभवाओ अपुधब्भूदो य अजुदसिद्धो य।
तह्मा दव्वगुणाणं अजुदा सिद्धित्ति णिद्दिट्ठा॥

samavattī sabhavāo apudhabbhūdo ya ajudasiddho ya ।
tahmā davvaguṇāṇaṁ ajudā siddhitti ṇiddiṭṭhā ॥

On account of the rise, subsidence, annihilation, mixed substance and annihilation of *karmas*, by the intrinsic nature of the Self unconditioned by *karmas*, *jīva* has five *bhāvas* (five characteristics). These are described in the Āgamas (scriptures).

Commentary

Karmas are considered to be physical in nature. These physical conditions determine the psychic characteristics. Five different classes of conditions are mentioned. Each of these is the causal condition determining its corresponding how about thought state in *jīva*.

These conditions are: (1) *udaya* (rise) of *karmas*; (2) *upaśama* (subsidence) of *karmas*; (3) *kṣaya* (eradication); (4) *kṣayopaśama* (mixed process of eradication and suppression); and (5) *pariṇāma* (unconditional thought).

These five conditions (four physical and one spiritual) determine respectively the *bhāvas*. And these are: (1) *audāyika bhāva*; (2) *aupaśamika bhāva*; (3) *kṣāyika bhāva*; (4) *kṣayopaśamika bhāva*; and (5) *pariṇāmika bhāva*.

The first one is the cause of bondage; the next three, i.e. two, three and four, are the causes of liberation. The last one is unconditioned by *karmas*. It is *nirupādhi* in character, whereas the other four are generated by the changes in physical conditions or *upādhis* (kārmic adjuncts and encumbrances). The last *pariṇāmika bhāva* is not causally connected with *saṁsāra* or *mokṣa*. It is *niṣkriya bhāva*.

57

वण्णरसगंधफासा परमाणुपरूविदा विसेसा हि।
दव्वादो य अणण्णा अण्णत्तपगासगा होंति॥

vaṇṇarasagaṅdhaphāsā paramāṇuparūvidā visesā hi ।
davvādo ya aṇaṇṇā aṇṇattapagāsagā hoṅti ॥

Being affected by the change in kārmic material, *jīva* experiences certain emotional, i.e. psychic, states. Whatever emotional state thus appears in consciousness is due to the direct causal agency of *jīva*. Thus say the Jaina scriptures.

Commentary by A. Chakravarti

Jīva is said to be the direct and immediate cause of the several emotional states brought about by kārmic materials. The extrinsic cause physical matter and the proximate cause is *jīva* itself. *Karmas* are always spoken of as of two kinds: *dravya karma* and *bhāva karma*. *Dravya karma* is distinctly physical. A peculiar combination of *paramāṇus* constitutes *karma pudgala* or kārmic matter. Matter under such modification has a special attraction towards *jīvas* ordinarily. *Jīva* in its worldly state is generally found in association with such kārmic matter. Kārmic matter may bring about a corresponding change in consciousness. This conscious change has a predominant affective tone. It is generally some form of emotion. This corresponding emotional or psychic state is *bhāva karma*. Since it is a psychic event, it is immediately determined by mind itself. It is this fact that is emphasized by Kundakunda in this verse.

Kundakunda recognizes two distinct causal agents which will be subsequently evident: *nimitta kartā* and *upādāna kartā* (indirect or extrinsic cause and substantial cause). *Jīva* is the substantial cause of all its modifications and matter may still be an extrinsic cause. Then *jīva* is the *upādāna kartā* of the *bhāvas* while kārmic matter is *nimitta kartā*.

58

दंसणणाणाणि तहा जीवणिवद्धाणि णण्णभूदाणि ।
ववदेसदो पुधत्तं कुव्वंति हि णो सभावादो ॥

daṁsaṇaṇāṇāṇi tahā jīvaṇivaddhāni ṇaṇṇabhūdāṇi ।
vavadesado pudhattaṁ kuvvaṁti hi ṇo sabhāvādo ॥

The different psychic *karmas* (*bhāva karmas*) such as coming into operation, rise (*udaya*) or fruition of material *karmas* (*aupadāyika*), the subsidential (*aupaśamika*), the destructive (*kṣāyika*) and partially destructive and partially subsidential (*kṣayopaśamika*) cannot happen in the consciousness of the Self without corresponding changes in physical *karmas* (*dravya karmas*). Therefore, the *bhāvas* (psychic states) like *audāyika*, which are brought about by the psychic *karmas*, may be said to be the effects of physical *karmas*.

Commentary

The physical *karmas* and psychic *karmas* are related as cause and effect, each of the other, in the capacity of indirect, extrinsic causal condition (*hetu*) or auxiliary cause (*nimitta kāraṇa*) of one another.

59

जीवा अणाइणिहणा संता णंता य जीवभावादो ।
सब्भावदो अणंता पंचग्गगुणप्पधाणा य ॥

jīvā aṇāiṇihaṇā santa ṇantā ya jīvabhāvādo ।
sabbhāvado aṇanta pañcaggaguṇappadhāṇā ya ॥

If the psychic states are really brought about or caused by the physical *karma*, how can the Self (*ātmā*) be said to be the cause of the psychic states? But the Self's causal agency is such that while giving up its own state, it can effect entirely alien or impure changes or impure mental states, which are also indirectly conditioned by physical *karmas*.

Commentary by Amṛtacandra

If the *bhāva* of the soul of the nature of *audāyika*, etc. is *karmakṛta*, i.e. engendered by *karma* (*dravya karma*), then the soul is not the causal agent (*kartā*) of *bhāvas* like *audāyika* and since non-doership (*akṛtatva*) is not desirable, it remains or is proved that the soul should be the causal agent (*kartā*) of *dravya karma*. But how can that be? From the internal self-referential point of view (*niścaya naya*), the Self does not do anything other (*anya*) by giving up one's own (*sva*) *bhāva*.

Commentary by Jayasena

Just as the soul is not the causal agent (*kartā*) of *bhāvas* like attachment (*bhāva karma*), from the pure *niścaya naya* if the soul is non-doer (*akartā*) (of attachment, etc. impure psychic states) from the pure *niścaya naya*, there will be the absence of bondage of *dravya karmas* with the result that there will be the absence of *saṁsāra*, i.e. the transmigratory existences of the cycle of births and deaths, then the soul will always be liberated (soul), which is evidently contrary to the reality of direct perception; this is the purport.

60

एवं सदो विणासो असदो जीवस्स होइ उप्पादो ।
इदि जिणवरेहिं भणिदं अण्णोण्णविरुद्धमविरुद्धं ॥

evaṁ sado viṇāso asado jīvassa hoi uppādo ।
idi jiṇavarehiṁ bhaṇidaṁ aṇṇoṇṇavirudhamaviruddhaṁ ॥

Physical *karma* (*dravya karma*) is the supportive factor (*nimitta*) or auxiliary, subsidiary or secondary cause (*nimitta-kāraṇa*) of *bhāva*, i.e. psychic or emotional states (*bhāva karmas*) of the soul, and *bhāva* (*karmas*), in turn, are the causal condition of (physical) *karmas*. There is, indeed, no *kartā* (causal agent or doer) of them (i.e. *dravya karma* and *bhāva karma*). Nevertheless, without the causal agent (*kartā*), the (physical and psychic) changes cannot happen.

Commentary by Amritachandra

Jīva bhāvas, i.e. the psychic states of the soul, are *karma-kṛta*, i.e. caused by *karmas* (*dravya karmas*) from the external, other-referential (*vyavahāra*) (point of view) and as *nimitta mātra*, i.e. only a supportive factor or auxiliary cause, from the internal, self-referential point of view (*niścaya naya*) (*audāyika*, etc.), neither *jīva bhāvas* are *karma-kṛta*, i.e. caused by *karma*, nor *jīva bhāvas* are the causal agents (*kartā*) of *karma*. And it is also evident that they (*jīva bhāvas* and *dravya karmas*) take place without the causal agent (*kartā*) because from *niścaya naya* the soul is the *kartā* of the states-of-evolution (*pariṇāma*) of *jīva*; and (material) *karmas* is the causal agent (*kartā*) of the states of evolution of *karma*.

Commentary by A. Chakravarti

Kundakunda here makes use of the distinction between *upādāna kartā* (substantive causal agent) and *nimitta kartā* (auxiliary or supportive causal agent or doer), i.e. intrinsic or substantive condition and extrinsic or alien condition. The soul is the essential, i.e. substantive cause of kārmic thought, the emotional states of desire, etc. Kārmic matter is the essential (substantive) cause of the changes in kārmic matter, i.e. the changes in each case form an independent series and yet the two series are corresponding and interrelated. *Bhāvas* (emotional states) is distinctly a psychic change. As a mental fact, it is immediately determined by the nature of consciousness. So is a change in kārmic matter. It is entirely determined by physical conditions. Matter cannot take the form of a mental state of the mind that undergoes a physical change.

Kundakunda seems to suggest a kind of psychophysical parallelism. But this parallelism is not merely the temporal correspondence of the two series. The parallelism is transcended and reconciled by the doctrine of *nimitta kartā*. Kundakunda has in his mind such a sharp distinction between the thinking thing and the unthinking thing as it is associated with Cartesianism and yet the two are related by a peculiar conception of a causal relation. The unthinking thing may be the *nimitta kartā* of the thinking thing and conversely the thinking thing may be the *nimitta kartā* of the other, though certainly one cannot be the substantive causal agent (*upādāna kartā*) of the other.

61

णेरइयतिरियमणुआ देवा इदि णामसंजुदा पयडी।
कुव्वंति सदो णासं असदो भावस्स उप्पादं॥

ṇeraiyatiriyamaṇuā devā idi ṇāmasaṅjudā payaḍī ।
kuvvaṁti sado ṇāsaṁ asado bhāvassa uppādaṁ ॥

The Self or the soul is the (substantive (*upādāna*)) causal agent or doer (*kartā*) of its own psychic states (*svakāsya-bhāvasya*);

it is not the *upādāna kāraṇa* of the changes in material *karma* (*pudgala karma*). Let the words of the Omniscient Lord (Jina) be understood thus.

Commentary by A. Chakravarti

The different states of existence are the different modifications of the *jīva* brought out by *upādhi* (kārmic conditions). Those *karmas* that determine the next individual state of *jīva* are *nāma karmas*. These *nāma karmas* are the real, causal conditions that lead the soul to manifest in a particular form. These forms appear and disappear. Birth and death are characteristics associated with these forms or modes. Just as waves may appear and disappear on the surface of the ocean as the result of winds, while the surface of water is really unaffected, so the *jīva*, remaining eternal and changeless in its real essence, is the basis of the different waves of existence appearing and disappearing. And these waves in the ocean of life are brought about by *upādhi*. The characteristics of birth and death, which are really true of the form of existence, are also predicated of the *jīva*.

62

उदयेण उवसमेण य खयेण दुहिं मिस्सिदेहिं परिणामे ।
जुत्ता ते जीवगुणा बहुसु य अत्थेसु विच्छिण्णा ॥

udayeṇa uvasameṇa ya khayeṇa duhiṁ missidehiṁ pariṇāme ।
juttā te jīvaguṇā bahusu ya atthesu vicchiṇṇā ॥

Kārmic matter itself, through its own essential nature, indeed brings about its own changes. Similarly, the *jīva* also through its own impure states of thought, that is conditioned by *karma*, brings about its own thought changes.

Commentary by A. Chakravarti

Kundakunda here emphasizes the fact that the mental states and the states of *karma* form two independent series. A change in the *jīva* is entirely determined by the antecedent mental change. Kārmic matter is determined by the antecedent physical change. Similarly, the mental change in the *jīva* is entirely determined by the antecedent mental change. In short, the *jīva* is the primary condition (substantive cause (*upādāna-kāraṇa*)) of mental changes in kārmic matter of its own changes. The physical changes of kārmic matter cannot be attributed to the *jīva* or the mental changes of the *jīva* to matter.

63

कम्मं वेदयमाणो जीवो भावं करेदि जारिसयं।
सो तेण तस्स कत्ता हवदित्तिय सासणे पढिदं॥

kammaṁ vedayamāṇo jīvo bhāvaṁ karedi jārisayaṁ ।
so teṇa tassa kattā havadittiya sāsaṇe paḍhidaṁ ॥

If the kārmic matter effects its own changes and similarly the soul brings about his own changes in himself, why should he enjoy the results or fruits of *karma*, for which he is not responsible and how can *karma* offer its fruits to him?

Commentary by Amr̥tacandra

If there is no causal agency or non-doership (*akartāpana*) of one another of *karma* and soul, then the fruit effected or done by other (*anya*) and experienced by other (*anya*) – relevance of such a question will arise.

The above question is answered in the next seven verses (*PKS* 64-70).

64

कम्मेण विणा उदयं जीवस्स ण विज्झदे उवसमं वा।
खइयं खओवसमियं तम्हा भावं तु कम्मकदं॥

kammeṇa viṇā udayaṁ jīvassa ṇa vijjhade uvasamaṁ vā ।
khaiyaṁ khaovasamiyaṁ tamhā bhāvaṁ tu kammmakadaṁ ॥

The matter bodies fill the entire world completely in every way, both subtle or minute and gross infinite of infinities is in number, and of multifarious kinds of constitution.

Commentary by Amr̥tacandra

Amr̥tacandra explains the principle of contiguity as follows. Just as the casket filled with collyrium powder becomes black by mere contact, so also the two distinct things, *jīva* (soul) and matter become related together.

65

भावो जदि कम्मकदो अत्ता कम्मस्स होदि किध कत्ता ।
ण कुणदि अत्ता किंचि विमुत्ता अण्णं सगं भावं ॥

bhāvo jadi kammakado attā kammassa hodi kidha kattā ।
ṇa kuṇadi attā kiṅci vimuttā aṇṇaṁ sagaṁ bhāvaṁ ॥

The Self by its own nature is the causal agent of its own impure psychic states like attachment, then at that same place, kārmic matter by its own nature gets inseparably bound with one another (*anyonya*), i.e. *jīva* manifests as modes of *karma* (such as knowledge-obscuring).

Commentary

Because of occupying of the same space, the kārmic matter, which as conditioned by the impure mental states, undergoes modifications. These modifications are really determined by matter itself, though the psychic states of *jīva* act as an indirect, extrinsic causal condition (*hetu*) or a supportive factor (*nimitta*) in that modification.

66

भावो कम्मणिमित्तो कम्मं पुण भावकारणं हवदि ।
ण दु तेसिं खलु कत्ताणि विणा भूदा दुकत्तारं ॥

bhāvo kammaṇimitto kammaṁ puṇa bhāvakāraṇaṁ havadi ।
ṇa du tesim khalu kattāṇi viṇā bhūdā dukattāraṁ ॥

Just as several molecular arrangements in matter are seen in diverse forms though are not caused by an alien agency, so also the manifestations in kārmic matter occur undetermined by alien cause. So do you understand.

Commentary by A. Chakravarti

Just as the mere presence of light from the sun or the moon is enough to create the fiery sunset or sunrise or the rainbow or the halo, these changes are all due to molecular arrangement in matter. The sunlight is not directly interfering with matter in producing changes. The changes are the necessary concomitants of the presence of light. Similarly, the presence of impure psychic states of attachment or aversion in *jīva* has as its concomitant, the changes in kārmic matter. The rainbow of several iridescent colours is nearly the concomitant of light and the different manifestations of kārmic matter are also the concomitants of mental states as attachment and aversion. Thus, while it is evident that there is no direct causal relation between the two series, yet one series is the concomitant of the other.

67

कुव्वं सगं सहावं अत्ता कत्ता सगस्स भावस्स ।
ण हि पोग्गलकम्माणं इदि जिणवयणं मुणेयव्वं ॥

kuvvaṁ sagaṁ sahāvaṁ attā kattā sagassa bhāvassa ।
ṇa hi poggalakammāṇaṁ idi jiṇavayaṇaṁ muṇeyavvaṁ ॥

Jīvas and kārmic materials are bound together strongly with one another, but when the time for the separation comes they fall apart, *karmas* offer the results of pleasure and pain, and *jīvas* enjoy them.

Commentary by A. Chakravarti

Jīva and matter in reality have only one causal function of generating their own respective modes and yet because of the concomitants they may be said to be interdetermining from the external, other-referential (*vyavahāra*) view. *Jīva*, because of its psychic states of attachment and aversion, develops an inclination towards matter. This inclination is only the *nimitta* (an external condition). Kārmic matter so determined gets bound to *jīva*. It is to be imagined that the material particles somehow cling to *jīva* and cloud its intrinsic radiance. Changes may occur in molecules as determined by temporal conditions. There may be aggregation or disintegration in them. When such changes take place in *dravya karma*, *jīva* experiences pleasure or pain. These are said to be offered by *karma* from the *vyavahāra* point of view, though as a matter of fact they are the modifications of *jīva*. Just as we experience the activity and the change in our [gross] body, so also we experience the change in our kārmic body, for both of them are physical. When we do not ordinarily question the bodily changes we are experiencing, we need not question the possibility of experiencing the changes of the kārmic body. Thus, ends the quality of *bhoktā* – the enjoyer. *Jīva* is shown to be the enjoyer.

68

कम्मं पि सगं कुव्वदि सेण सहावेण सम्ममप्पाणं।
जीवो वि य तारिसओ कम्मसहावेण भावेण॥

kammaṁ pi sagaṁ kuvvadi seṇa sahāveṇa samma-
mapāṇaṁ ।
jīvo vi ya tārisao kammasahāveṇa bhāveṇa ॥

According to the internal, self-referential point of view (*niścaya naya*), physical *karma* is the causal agent of its own modifications. But from the external, other referential point of view (*vyavahāra naya*), it is the causal agent to doer (*kartā*) of the psychic states (*jīva bhāvas*) in *jīva*. In other words, from the *niścaya naya*, it is the *kartā* of its own modes and from *vyavahāra*, it is the *kartā* of *jīva bhāva*, but it is not the experiencer or enjoyer. *Jīva* alone is the enjoyer of the kārmic effects or fruits of *karma* (*karma-phala*) due to its nature of consciousness.

Commentary by Amṛtacandra

Thus, it is decided that from the *niścaya naya*, *karma* is the causal agent or the doer (*kartā*) of its own modes or changes; from *vyavahāra naya*, it is the *kartā* of *jīva bhāva*; *jīva*, also from the *niścaya naya*, is the *kartā* of its own psychic states and from *vyavahāra*, it is the *kartā* of kārmic matter.

Therefore, from both points of view, *karma* is the *kartā*, in the same way it is not the enjoyer from any one point of view. For what reason? It is because, it does not have the good disposition (*sadbhāva*) of experiencing consciousness. Therefore, due to consciousness, the *jīva* alone is famous for being the experiencer or enjoyer of *karma-phala* (fruits of *karma*) – in some respect of the Self's states-of-evolution of pleasure and pain, and in some respect of the desired and undesired sense objects (*viṣaya*).

69

कम्मं कम्मं कुव्वदि जदि सो अप्पा करेदि अप्पाणं।
किध तस्स फलं भुंजदि अप्पा कम्मं च देदि फलं॥

kammaṁ kammaṁ kuvvadi jadi so appā karedi appāṇaṁ ।
kidha tassa phalaṁ bhuṅjadi appā kammaṁ ca dedi phalaṁ ॥

The soul, thus, the causal agent of its own *karma*, is an enjoyer of the fruits thereof and covered by infatuation wanders or roams about in this world of transmigratory existences of births and deaths (*saṁsāra*), which is limited for the faithful and unlimited for the unfaithful.

Commentary by A. Chakravarti

Jīva is the Lord of his own career, because he effects his own *karma*, he enjoys the fruits thereof. He may roam about in the world of *saṁsāra* as conditioned by his own actions. He may finally liberate himself from the bondage by walking the path of three jewels, i.e. enlightened world-view or right belief, enlightened knowledge and enlightened conduct. Then he becomes free from the kārmic adjuncts and encumbrances (*karma upādhi*).

70

ओगाढगाढणिचिदो पोग्गलकाययेहिं सव्वदो लोगो ।
सुहमेहिं वादरेहिं य णंताणंतेहिं विविहेहिं ॥

ogāḍhagāḍhaṇicido poggalakāyayehiṁ savvado logo ।
suhamehiṁ vādarehiṁ ya ṇaṅtāṇaṁtehiṁ vivihehiṁ ॥

After obtaining the path (of triple jewels) revealed by the Omniscient Lord (Jina) and due to the subsidence, destruction or partly destructive and partly subsidential of the deluded view, the resolute or preserving person wanders or behaves on the path in accordance with the enlightened knowledge, then he obtains a liberated state.

71-72

अत्ता कुणदि सहावं तत्थ गदा पोग्गला सभावेहिं ।
गच्छंति कम्मभावं अण्णोण्णागाहमवगाढा ॥

attā kuṇadi sahāvaṁ tattha gadā poggalā sabhāvehiṁ ।
gacchaṅti kammabhāvaṁ aṇṇoṇṇāgāhamavagāḍhā ॥

जह पुग्गलदव्वाणं पहुप्पयारेहिं खंधणिव्वत्ति ।
अकदा परेहिं दिट्ठा तह कम्माणं वियाणाहि ॥

jaha puggaladavvāṇaṁ pahuppayārehiṁ khaṅdhanivvatti ।
akadā parehiṁ diṭṭhā taha kammāṇaṁ viyāṇāhi ॥

The great soul, characterized by *upayoga*, i.e. conscious attentiveness, function or manifestation of consciousness, is essentially one alone. It is again of two kinds, has three fundamental qualities, roams about in four conditions of mundane existence (*gati*); and is marked by five primary emotional or psychic states. He moves about in the world along the six directions; is capable of admitting sevenfold predications. He has eight qualities, generates the nine categories (*padārthas*) or basic principles of life (*tattvas*) and manifests in ten states of existence; thus, he is said to be.

Commentary

The great soul is (1) one alone due to being characterized by *upayoga*; (2) due to having institution (*darśana*) and knowledge, it is of two kinds; (3) due to having three aspects of consciousness, viz. (a) *karma-phala-cetanā*, i.e. consciousness of experiencing merely the fruits of one's past *karma*; (b) *karma-cetanā*, i.e. consciousness of one's activities; and (c) *jñāna-cetanā*, i.e. knowledge consciousness; or of the substance (*dravya*) being characterized by origination (*utpāda*), decay (*vyaya*) and permanency (*dhrauvya*) [or three jewels of *darśana*, *jñāna* and *caritra*]; (4) roams about in four conditions of existence (*gati*), viz. (i) human, (ii) hellish, (iii) celestial, and (iv) animal, etc. (*tiryañca*); (5) five primary psychic states (*TS* II.1), viz. (a) *audāyika*, i.e. coming into operation or rise (*udaya*) of *karmas*; (b) *aupaśamika*, i.e. subsidential; (c) *kṣāyika*, i.e. destructive; (d) *kṣayopaśamika*, i.e. partly destructive and partly subsidential; and (e) *pariṇāmika bhāva*, the innate pure natural disposition of the Self, i.e. consciousness-as-such, which is non-relational to or independent of *karmas* (*karma-nirapekṣa*); (6) six directions of the world for the possible movements of *jīva*; (7) sevenfold predications, expressions or judgements, i.e. in certain respect a thing is (*syāt asti*), is not (*syāt nāsti*), etc.; (8) knowledge-obscuring, etc. eight kinds of *karma* that the mundane soul has or, the eight infinite qualities that the perfect *jīva* has, viz. *ananta* (infinite) *jñāna*, *ananta darśana*, *ananta vīrya* (vigour), *ananta sukha* (bliss), etc.; (9) nine categories or principles (*tattvas*) – *jīva* (conscious being), *ajīva* (non-conscious entity and inert matter), virtue or merit (*puṇya*), sin or vice (*pāpa*), influx (*āśrava*), stoppage (*saṁvara*) of the influx, shedding (*nirjara*), and bondage (*bandha*) of *karmas* and liberation (*mokṣa*); and (10) ten states of existence. These are (i) *pṛthvīkāyika* (earth-bodied), (ii) *āpakāyika* (acquatic or water-bodied), (iii) *tejakāyika* (fire borne), (iv) *vāyukāyika* (air borne), (v) *saddharma vanaspati* (a vegetable with many souls in one body) (*saddharma* is so-called because the many souls are the same with respect to their bodies, food, respiration and age), (vi) *pratyeka vanaspati* (one body-one-souled vegetables), (vii) two-sensed, (viii) three-sensed, (ix) four-sensed, and (x) five-sensed ten places in which the *jīva* exists.

73

जीवा पुग्गलकाया अण्णोण्णागाढगहणपडिबद्धा ।
काले विजुज्जमाणा सुहदुक्खं दिंति भुंजंति ॥

jīvā puggalakāyā aṇṇoṇṇāgāḍhagahaṇapaḍibaddhā ।
kāle vijujjamāṇā suhadukkhaṁ diṁti bhuṁjaṁti ॥

There are four aspects of kārmic bondage, viz. (1) *prakṛti* (nature, quality or class of *karmas* that is bound); (2) *sthiti* (duration); (3) *anubhāga* (intensity of fruition of *karmas*); and (4) *pradeśa bandha* (quantity or mass of material particles assimilated). Then, there are liberated souls, entirely liberated from kārmic bondage, which go upwards till they reach the summit of the upper world. Other mundane souls move about in all the directions except the four diagonal corners.

Commentary by A. Chakravarti

The *jīva* and *dravya* (matter) in reality have only one causal function of generating their own respective modes and because of the concomitants they may be said to be interdetermining from the *vyavahāra* view. Because of its emotional states of desire and hatred, the *jiva* develops an inclination towards matter. This inclination is only the *nimitta* (external) condition. Kārmic matter so determined gets bound to the *jīva*. It is to be imagined that the material particles somehow cling to the *jīva* and cloud its intrinsic radiance. Changes may occur in molecules as determined by temporal conditions. There may be an aggregation or disintegration in them. When such changes take place in *dravya karmas*, the *jīva* experiences pleasure or pain. These are said to be offered by *karmas* from a relative point of view though, as a matter of fact, they are the modifications of the *jīva*. One of the commentators says that just as we experience the activity and the change in our body, so also, we experience the changes in our kārmic body for both of them are physical. When we do not ordinarily question our experiencing bodily changes, we need not question the possibility of experiencing the changes of the kārmic body. Thus, ends the quality of *bhoktā* – the enjoyer *jīva* is shown to be the enjoyer.

CHAPTER TWO

Pudgala Dravya Astikāya

74

तम्हा कम्मं कत्ता भावेण हि संजुदोध जीवस्स ।
भोत्ता दु हवदि जीवो चेदगभावेण कम्मफलं ॥

tamhā kammaṁ kattā bhāveṇa hi saṅjudodha jīvassa ।
bhottā du havadi jīvo cedagabhāveṇa kammaphalaṁ ॥

Be it understood that matter (*pudgala*) exists in four main modes: *skandhas*, *skandhadeśas*, *skandhapradeśas* and primary or minutest indivisible atoms (*paramāṇus*). Out of these four basic modifications, multifarious modes of matter are formed.

Commentary by Amr̥tacandra (AC)

Pudgala substance (*dravya*) at some time (*kadācita*) are (1) from *skandhas* mode, which are the aggregates of atoms and refers to complete molecular constitution; (2) *kadācita* from *skandhadeśa* mode, which are said to be incomplete but an aggregate nevertheless; and (3) *kadācita* from *skandhapradeśa* mode. These three substances have a different molecular constitution. The last class (4), i.e. the *paramāṇu*, is the unit constituting the other three classes or modes. In this way, there are four kinds of *pudgala dravya*.

75

एवं कत्ता भोत्ता हीज्झं अप्पा सगेहिं कम्मेहिं ।
हिंडति पारमपारं संसारं मोहसंछण्णो ॥

evaṁ kattā bhottā hījjhaṁ appā sagehiṁ kammehiṁ ।
hiṇḍati pāramapāraṁ saṁsāraṁ mohasañchaṇṇo ॥

The complete molecule of matter is *skandhas*: one half of it is *skandhadeśa*; one half of that half is *skandhapradeśa*: and what cannot be divided is *paramāṇu* (minutest primary atom). Thus, say the learned.

Commentary by Amr̥tacandra (AC)

An aggregate of infinite number of *paramāṇu* is in *skandhas* mode. Beginning from two *paramāṇus*, three, four, etc. more *paramāṇus* will generate infinite modes of the nature of *skandhas*.

76

उवसंतखीणमोहो मग्गं जिणभासिदेण समुवगदो ।
णाणाणुमग्गचारी णिव्वाणपुरं वजदि धीरो ॥

uvasaṅtakhīṇamoho maggaṁ jiṇabhāsideṇa samuvagado ।
ṇāṇāṇumaggacārī ṇivvāṇapuraṁ vajadi dhīro ॥

Skandhas are of two kinds: (1) gross, that can be perceived by the senses; and (2) subtle which are beyond sense perception. They are also called matter conventionally (from *vyavahāra*). They manifest themselves in six different modes which completely fill the three worlds.

Commentary by Amr̥tacandra (AC)

The term *pudgala* is so-called as it is filling up (*pūrṇa*) and decaying (*galana*), especially applicable to the *paramāṇu*. Those are the indestructible material bases of the world. The *skandhas* (molecules), though derivative and secondary, are also called *pudgala*. These *skandhas* have the characteristic of touch, taste, smell, sound and colour. They can increase or decrease since they are aggregates. They can grow and decay. These *skandhas* (molecular aggregates) are of six different forms which completely fill the three worlds.

These are:

1. Gross–gross (*badara–badara*): Solid that which cannot become combined or one, when broken, like wood or stone.
2. Gross (*badara*, liquid): One whose parts become one again when broken, as milk, *ghee*, oil, water, etc.
3. Gross–subtle (*badara–sūkṣma*): The larger cannot be broken or divided or caught as shadow, lightning or darkness.
4. Subtle–gross (*sūkṣma–badara*): Though subtle or minute, yet evident to the senses of taste, smell, colour, etc.
5. Subtle (*sūkṣma*): Minute and imperceptible as kārmic matter.
6. Subtle–subtle (*sūkṣma–sūkṣma*): Those below kārmic matter up to *skandhas* made up of two atoms.

Commentary by Jayasena

The soul of the nature of the liberated Self lives by pure conscious vitality is from internal, self-referential point of view (*niścaya naya*). From external, other-referential point of view (*vyavahāra naya*), the soul lives by physical vitalities of five senses, three energies of mind, speech and body, duration of life and respiration.

That which are filled up and decay (*pūrṇa–galana*) through the state-of-evolution of touch, taste, smell and colour, these four qualities continue to grow and decay in their portions, such *paramāṇu* are called *pudgala*-like *skandhas*. From *vyavahāra naya*, from the *skandhas* of two *paramāṇus* to the aggregates of infinite *paramāṇu*, gross and subtle, are also called *pudgala*. These are like *padārthas* (things) are seen, that alone is called *loka* (universe). This *loka* is formed by *pudgala* (matter) like six substances (*dravyas*). Neither any particular or special person has created it, nor anyone can destroy it, nor anyone has maintained or preserved it.

77

एको चेव महप्पा सो दुवियप्पो त्तिलक्खणो होदि।
चदु चंकमणो भणिदो पंचग्गगुणप्पधाणो य॥

eko ceva mahappā so duviyappo ttilakkhaṇo hodi ।
cadu caṅkamaṇo bhaṇido pañcaggaguṇappadhāṇo ya ॥

The last limit of all *skandhas* is the *paramāṇu*. The same is eternal, unsounding (silent), occupying one space-point and of corporeal form.

Commentary by Amṛtacandra

This verse explains a *paramāṇu*. It is the smallest portion and it cannot be further subdivided (*avibhāgī*). Since it occupies one space point, it is one (*eka*). Since it is the eternal and inalienable substratum of corporeal things, it is permanent (*nitya*). It is without beginning and end. Since it is the constitutive basis of physical objects, it is the generator of physical things. Since it cannot by itself bring about sound, it is non-sounding or silent. Sound is not the characteristic of *paramāṇu*. Sound is of the nature of mode (*paryāya*) of *pudgala skandhas*, as described in *Pañcāstikāyasāra* 79.

78

छक्कापक्कमजुत्तो उवउत्तो सत्तभङ्गसब्भावो ।
अट्ठासओ णवत्थो जीवो दसट्ठाणगो भणिदो ॥

chakkāpakkamajutto uvautto sattabhaṅgasabbhāvo ।
aṭṭhāsao ṇavattho jīvo dasaṭṭhāṇago bhaṇido ॥

That which is corporeal (*mūrta*) only by instruction and is the essential cause of the four primary substances [viz. earth, water, fire and air], that is to be understood as *paramāṇu*. It is characterized by the quality of states-of-evolution and is in itself *aśabda* (unsounding).

Commentary

Due to being corporeal, *paramāṇu* is characterized by sense qualities of touch, taste, smell and colour. It is the cause of the states-of-evolution of four primary substances of earth, water, fire and air.

Commentary by A. Chakravarti

Wonderful are the qualities of primary atoms when their potential characteristics are taken into consideration. The non-manifested sense qualities become actually an aggregate (*skandhas*). But in case of sound it cannot be said that it is present in the atom [*paramāṇu*] even in *śakti* (potential) for sound implies several molecules of several spatial units. Therefore, such a characteristic cannot consistently belong to an atom [*paramāṇu*] which has only one spatial unit.[1]

79

पयडिट्ठिदिअणुभागप्पदेसबंधेहिं सव्वदो मुक्को।
उड्ढं गच्छदि सेसा विदिसावज्जं गदिं जंति॥

payaḍiṭṭhidiaṇubhāgappadesabaṁdhehiṁ savvado mukko ।
uḍḍaṁ gacchadi sesā vidisāvajjaṁ gadiṁ jaṁti ॥

Sound is generated by *skandhas*, i.e. *skandhas* are the causal agency of sound. And *skandhas* are the aggregates of *paramāṇu*. When these aggregates strike one another, sound is produced, which may be natural or artificial (hence sound may be indirectly associated with *paramāṇu* also).

Commentary by Amṛtacandra

In reality by nature, *śabda* (sound) is the mode (*paryāya*) of the nature of one *skandha*, which is constituted of infinite *paramāṇu*.

Commentary by A. Chakravarti

If these *skandhas* strike of their own accord, then there is a natural (*svābhāvika*) sound. If the striking is due to other agencies, then

[1] The English translator (A. Chakravarti) of the *Pañcāstikāyasāra* and many others often use the term "atom" for *paramāṇu*, which is not correct in view of the latest advances in science. An atom (*aṇu*) is now divisible into electrons, protons and neutrons, and hence the translation of *paramāṇu* as atom is not proper.

it is *prāyogika* (purposeful) sound. The illustrations of *svābhāvika* (natural) sound of thunder of the clouds and the roar of the sea. *Prāyogika* (purposeful) sound is again of two kinds: *bhāṣātmaka* (pertaining to language) and *abhāṣātmaka* (non-language). The language sound may be *akṣarātmaka* or *anākṣarātmaka* [as the divine sound, echo or tone (*divya vāṇī* of the Omniscient Lord (Kevalī) (*PKS* 79 JS)], articulate and inarticulate. The articulate sound is made up of an alphabetical sound; the inarticulate is the language of animals.

Non-language sounds are of four kinds:

1. *Tata*, sound produced by stringed instruments, such as *vīṇā* or *sitār*.
2. *Vitata*, the sound produced by a drum.
3. *Ghana*, sound produced by cymbals, etc.
4. *Suṣira*, sound produced by pipes and other wind instruments (Chakravarti, based on *PKS* 79 JS).

80

खंधा य खंधदेसा खंधपदेसा य होंति परमाणू।
इति ते चदुव्वियप्पा पुग्गलकाया मुणेयव्वा॥

khaṅdhā ya khaṅdhadesā khaṅdhapadesā ya hoṅti para-māṇū ।
iti te caduvviyappā puggalakāyā muṇeyavvā ॥

A *paramāṇu* is eternal, it is special and yet non-special; it is the differentiating factor of *skandhas* and is also the cause of their changes. It is also the determinant of time in number.

Commentary by A. Chakravarti

Paramāṇu is associated with one space-point. Since it is not capable of disintegration, it is indestructible and eternal. Though it occupies only one space-point since it is the substratum of colour and other sense qualities, it has spaciality accommodating qualities. But as it has only one space-point and as it is without beginning, middle or end, and as it cannot accommodate another space-point, it may also be called non-spatial. Being the constituent element of *skandhas*, it may be said to be the determinant of the difference of *skandhas*. For the same reason, it is the substantial cause of *skandhas*. By its own change of position, it becomes a measure of time. A single instant of time corresponds to a single shift of a *paramāṇu* from one position to the immediately next. It is also the measure of number or quantity. Being the constitutive element of *skandhas*, it brings about a quantitative difference of things (*dravya-saṅkhyā*). Since its associated space point is the constitutive element of space, it is indirectly the cause of a quantitative difference of space (*kṣetra-saṅkhyā*). Since its motion from point-to-point corresponds to the duration of time, it is also the basis of quantitative difference of time (*kāla-saṅkhyā*). Again, because it is the basis of modification of things through aggregation and disintegration, it is also the condition of the quantitative difference of modes (*bhāva-saṅkhyā*).

Thus, according to the Jaina view, the *paramāṇu* is the direct unit of things and the indirect unit of space, time and change. The quantitative difference in these things and also the qualitative difference in physical objects may ultimately be traced to the constitutive of *paramāṇu*.

81

खंधं सयलसमत्थं तस्स दु अद्धं भणंति देसोत्ति।
अद्धद्धं च पदेशो परमाणु चेव अविभागी॥

khaṁdhaṁ sayalasamatthaṁ tassa du addhaṁ bhaṇaṁti desotti ।
addhaddhaṁ ca padeśo paramāṇu ceva avibhāgī ॥

The *paramāṇu* at one time has only a single (one) taste, one colour and one smell, and two touch. It is the cause of sound, though itself unsounding; and though it is within the *skandhas* due to constituting them, it is definitely only one substance (*dravya*) alone.

82

वादरसुहुमगदाणं खंधाणं पुग्गलोत्ति ववहारो ।
ते होंति छप्पयारा तेलोक्कं जेहिं णिप्पण्णं ॥

vādarasuhumagadāṇaṁ khaṅdhāṇaṁ puggalotti vavahāro ।
te hoṅti chappayārā telokkaṁ jehiṁ ṇippaṇṇaṁ ॥

Defining matter (*pudgala*), the verse states: "Whatever is perceived by the senses, the sense organs, the various kinds of bodies of *jīvas*, the physical mind or brain, the *karmas* and all that is corporeal (*mūrta*) objects. Understand that all these are *pudgalas* (matter)."

Commentary by A. Chakravarti

In this verse, the author enumerates the different states of matter. All objects of sense perception are corporeal. An object may be apprehended through any one of the five senses. Therefore, anything that has a sense quality of colour, touch, smell, taste or sound comes under this definition. Not only the objects of sense perception but also the sense organs are physical. Third, the different bodies of *jīvas* are also physical. These are five in number: *audārika*, *vaikriyika*, *āhārika*, *taijasa* and *kārman*. The author includes *dravya-manas* (physical mind) under the same class. Evidently, it means the brain which is, no doubt, physical. Physical (*dravya*) *karmas* are also material since they are constituted by physical *paramāṇu*. And again he implies no-*karma* (physical body) *pudgalas*, organized matter forming part of an organism. This no-*karma* matter refers to matter assimilated by the organism through the process of metabolism. And by the word "others" he refers to the several *paramāṇus* and *skandhas* not already enumerated. All these are material.

The five kinds of bodies (*śarīras*) are as follows:

1. *Audārika-śarīra*, i.e. *sthūla* (gross). The physical body of all men and animals (*TS* II.36). It is called *audārika* because it is constituted by gross matter. It is born from the *udara* (womb). The *audārika-śarīra* refers to organic bodies of animals and humans.
2. *Vaikriyika-śarīra*, the body of hellish and celestial beings, which can change at will (*TS* II.36). The term *vikriya* implies the wonderful body transformations that are associated with a celestial being (*deva*). A *deva* can assume any kind of body he pleases from minute to huge forms. It is a physical body of a fluid nature, like that of celestial or hellish beings.
3. *Āhārika-śarīra*, i.e. (assimilative) (the spiritual man like emanation from the head of a saint in doubt, *TS* II.36). When a *yogī* during *tapas* (austerities) has certain difficulties about intricate facts of reality, there shoots forth a kind of subtle body from him. This stretches out so far as to be in communion with another well-informed person, from whom the information sought for is secured by a sort of a telepathic response. The shooting body of the *yogī* is *āhārika-śarīra.*
4. *Taijasa-śarīra*, i.e. electric (body of mundane souls formed of fine molecules of electric matter called *taijasa varganā* (*TS* II.36)). This refers to the shining body or the light body, i.e. the fine material molecules of electric matter called *taijasa varganā*, which is the electric body.
5. *Kārman-śarīra,* body made of fine molecules of kārmic matter called *kārman vargaṇā* (*TS* II 36), subtle body, i.e. this refers to the kārmic body of each *jīva*. These are made up of molecules (*vargaṇā*) of subtle matter, called *kārman vargaṇā* which fill the universe, of which the kārmic body is made.

The first (*audārika* body) alone is the object of sense perception and the others are subtle bodies. The succeeding one is subtler than the preceding one in order.

(*The following is from Tattvārthasūtra by J.L. Jaini*)

II.36: The first three bodies are made of *āhārika varganā*, molecules of assimilatative matter.

II.37: Of these five bodies, each successive one (is) fine, i.e. subtler than the one preceding it.

II.38: From the first of the third body, each one has innumerable times the number of *paramāṇus* (which are in the one preceding it).

II.39: Of the last two, i.e. the electric body and the kārmic bodies, each one compared with the body immediately preceding it has an infinitefold (number of *paramāṇus*).

II.40: The electric and kārmic bodies are unpreventable (in their passage, i.e. they can penetrate and permeate up to the end of the (universe).

II.41: The fluid (*vaikriyika*) and *āhārika* assimilative bodies can also penetṛate but at some places they cannot. Their penetration does not extend beyond *trasa nāḍī*, i.e. the mobile channel beyond where mobile souls are not found.

II.41: And their connection (i.e. of the electric and the kārmic bodies with the soul is) without beginning. That is, these two are always found in every soul and are absent only in liberated souls (*siddha*). But, of course, being material, they are always discarding the old constituent particles and attracting and assimilating new particles in the place of the old. From this point of view, a particular electric or kārmic body is said to have a beginning.

II.42: (The electric and the kārmic bodies are always found) with all (mundane souls).

II.43: Along with these two, i.e. the electric in the kārmic bodies, in their distribution at one and at the same time, with one soul (there can be utmost up to four, i.e. these two in one or two more bodies), i.e. a soul can never have all the five bodies at once. Electric and kārmic bodies it must always have. But if it has more, it can add only one or two more as follows:

- In *vigrahati* or passage from one incarnation to another, the soul is only two bodies, the *tejas* (electric) and the kārmic.
- In the human and subhuman condition of existence, the soul has three, i.e. the electric, the kārmic and the physical bodies.
- But the saints sometimes have these three and the *āhārika* (assimilative) body also, i.e. four bodies at one and the same time. This is the highest limit.
- The hellish and celestial beings also have three bodies, viz. the electric, the kārmic and the fluid bodies.

II.44: The last body, i.e. the kārmic, is *nirupabhoga.* It cannot be the means of enjoyment of the soul through the senses and the mind, as the physical body can be; for example, the kārmic body bears no sound sense, no sight, etc.

Note: The electric body, like the kārmic body, *nirupabhoga*. There is no *yoga* on vibration or cause of attraction of subtle matter in it.

II.45: The first, i.e. the physical body is produced along with the electric and the kārmic bodies, in beings who are born in the embryonic way (*garbha*, i.e. pregnancy) or by spontaneous generation (*sammūrcchanā*).

II.46: The fluid body is found along with the electric in the kārmic bodies in those who are born by instantaneous rise.

II.47: And fluid body can also be attained by other cause, i.e. by *labdhi* (attainment) due to special austerities. But this is, of course, in exceptional circumstances.

II.48: Even the electric body can be the product of special austerities. This occurs in two ways:

1. *Śubha taijasa*: Beneficient electric body, e.g. a saint with supernatural powers sees famine or plague, etc. in a country and is moved to compassion. His austerities enable his electric body to overflow itself, and issuing out of his right shoulder go and to remove the causes of famine, plague, etc. and then

come back and be reabsorbed in the same way in which it went out.

2. *Aśubha taijasa*: Maleficient electric body, e.g. the same saint being moved to anger, the duplicate body will go forth from the left shoulder and having consumed the object of anger, will return to go into the saint and consume him as well.

II.49: And the *āhāraka* body (is) beneficial pure and impreventible and found only in a *pramatta saṁyata* saint (i.e. one in the sixth stage of spiritual development with imperfect vows). It is (1) *śubha* (always) beneficient; (2) *viśuddha*, pure (the production of meritorious *karmas*); and (3) *avyāghātī*, unpreventable (by anything in its course).

The *āhāraka* body is the spiritual man like emanation that flames forth from the head of a saint when he wants to remove his doubt on some momentous and urgent point, the flaming form flashes across space, touches or sees the feet on some Omniscient Lord (Kevalī) or a saint with perfect scriptural knowledge (*śruta-kevalī*) like Bhadrabāhu, the preceptor of Candragupta Maurya, and thus having removed the doubt re-enters the saint's brain.

All this is done, of course, with lightning rapidity, but it never takes more than one *antaramuhūrtta*, i.e. 48 minutes, minus one instant.

II.50: The hellish beings (and those who are) *sammūrcchanā* spontaneously generated (are of) a common or neuter sex.

II.51: The celestial beings never have a common sex. They are always masculine or feminine.

II.52: The remaining (beings, i.e. those born of an embryo, have) three sexes (i.e. they can be masculine, feminine or common). *Liṅga* (sex) is of two kinds: *dravya-liṅga* (objective sex, differentiating organs of the sexes); and *bhāva-liṅga* (subjective sex, thought activity relating to sexual inclinations).

II.53: (Those who are) born by instantaneous rise, i.e. hellish and celestial beings (those who are in their) last (incarnation *carama* body) with the highest (kind of) physical body (and those whose) age is innumerable years (e.g. human and subhuman beings in *bhoga-bhūmi* (land of enjoyment), i.e. in a condition of life where there is all enjoyment and no labour like agriculture, etc. (these three live the full span of) their life. (It can) never (be) cut short (by themselves or others).

The remaining beings, viz. human and subhuman beings in *karma-bhūmi* (land of work), i.e. the condition of life where agriculture, etc. are necessary for sustenance, can have their lives cut short, by external causes, as accident, suicide, disease, etc., etc.

Thus ends the chapter on *Pudgala Astikāya.*

CHAPTER THREE

Dharma and Adharma

83

पुढवि जलं च छाया चउरिंदियविसयकम्मपाओग्गा ।
कम्मातीदा येवं छब्भेया पोग्गला होंति ॥

puḍhavi jalaṁ ca chāyā cauriṅdiyavisayakammapāoggā ।
kammātīdā yevaṁ chabbheyā poggalā hoṅti ॥

Dharmāstikāya is devoid of qualities of taste, colour, smell, sound and touch. It pervades the whole world. It is indivisible or continuous because of inseparability; it is extensive because of its coextensiveness with space. Though in reality of one space-point (*pradeśa*), yet in *vyavahāra* (external, other-referential) aspect, it has innumerable *pradeśī*.

Commentary by Amr̥tacandra

Dharmāstikāya and *adharmāstikāya* are distinctly peculiar to the Jaina system of thought. The former is the principal or medium of motion and the latter is the principle of rest. These terms are used in a technical manner by Jaina writers. Many non-Jaina scholars, both European and Indian, have misunderstood these technical names.

In this verse, the author describes the nature of *dharma*, the principle of motion. Since it has no sense qualities, it is of non-corporeal nature (*amūrta svabhāva)*, i.e. it is not physical. The sense qualities are generally associated with matter (*pudgala*).

The influence of *dharmāstikāya* is coextensive with the whole world. Therefore, it may be said to be coextensive with *lokākāśa* without any gaps or intervals. Due to being *ayutsiddha*, i.e. not having joining together (*asaṁyogī*) of *pradeśa*, it is indivisible or continuous (*akhaṇḍa*). It is by nature coeval with the world and coextensive with space; and because of the latter character, it is an *astikāya*. Even though being of one (*eka*) *pradeśī* from internal, self-referential point of view (*niścaya naya*), it is of innumerable *pradeśī* from the external, other-referential point of view (*vyavahāra naya*).

84

सव्वेसिं खंधाणं जो अंतो तं वियाण परमाणू।
सो सस्सदो असद्दो एक्को अविभागि मुत्तिभवो॥

savvesiṁ khaṁdhāṇaṁ jo aṁto taṁ viyāṇa paramaṇū ।
so sassado asaddo ekko avibhāgi muttibhavo ॥

Due to its *agurulaghu* quality (see *PKS* 24), the *dharmāstikāya* has infinite manifestations of the non-corporeal nature, it has indeed real existence, i.e. *nitya* (permanent). Itself being unaffected by moment, it acts as a supportive factor or auxiliary cause (*nimitta kāraṇa*) of motion of those (substances, i.e. matter and *jīva*), that can move.

Commentary by Amṛtacandra

Because of its *agurulaghu* quality (see *PKS* 24), the *dharmāstikāya* is always evolving due to its nature of persistence through origination and decay, i.e. maintaining its nature, hence it is *nitya* (permanent). Though in itself being apathetic, inert or passive, indivisible and of being only supportive factor, it is the *nimitta kāraṇa* in the movement of *jīva* and matter (*pudgala*), (only two of the five *astikāyas* and six substances), that can move.

85

आदेशमत्तमुत्तो धादुचदुक्कस्स कारणं जो दु ।
सो णेओ परमाणू परिणामगुणो सयमसद्दो ॥

ādeśamattamutto dhāducadukkassa kāraṇaṁ jo du ।
so ṇeo paramāṇū pariṇāmaguṇo sayamasaddo ॥

Just as water (in itself being apathetic or indifferent) favours, supports or assists in the movement of fish, so also *dharma dravya* (in itself indifferent) supports or assists in the motion of *jīvas* and *pudgalas*.

Commentary by A. Chakravarti

Kundakunda explains the function of *dharmāstikāya* by a very striking example. Water is the indispensable condition for the life of the fish. The fish live and move only in water. But water neither moves with the moving fish nor does it stimulate the fish to move. It is practically indifferent towards the moving fish. If the fish moves, it is due to its own intrinsic and spontaneous activity, to the causal agency of water. Such is the relation of *dharmāstikāya* to objects of the world. If objects move from one place to another, the movement is due to the intrinsic condition of the object. But still *dharmāstikāya* is the *sine qua non* of motion of the object of the world. In short, it is merely the condition and not the generative cause, of motion of *jīva* and matter (*pudgala*).

86

सद्दो खंधप्पभवो खंधो परमाणुसंगसंघादो ।
पुट्ठेसु तेसु जायदि सद्दो उप्पादगो णियदो ॥

saddo khaṅdhappabhavo khaṅdho paramāṇusaṅga-saṅghādo ।
puṭṭhesu tesu jāyadi saddo uppādago ṇiyado ॥

The nature of *adharma* substance is similar to that of *dharma* substance; know that. But it is like the earth (which is the resting place of things) the supportive factor or the causal condition of rest for things moving on their own (i.e. both *jīva* and *pudgala*).

Commentary by A. Chakravarti

Adharma (principle of rest) has all the characteristics associated with *dharma* (principle of motion). This is also devoid of sense qualities. This is also non-corporeal. This is in itself a non-spatial and yet it is coexpensive with *lokākāśa*. These characteristics are in common with the principle of motion. But it has its own differentiating quality. In this respect, it is compared with earth which is the resting place of things. Moving things, whether animate or inanimate, are not arrested and brought to rest by the earth. But if there is no earth to support, there will be no possibility of rest for the moving things. Similarly, the *adharmāstikāya* without interfering with motion itself is the indifferent/passive or the cooperative cause, of rest for the moving things.

87

णिच्चो णाणवकासो ण सावकासो पदेसदो भेत्ता ।
खंधाणं पि य कत्ता पविहत्ता कालसंखाणं ॥

ṇicco ṇāṇavakāso ṇa sāvakāso padesado bhettā ।
khaṅdhāṇaṁ pi ya kattā pavihattā kālasaṅkhāṇaṁ ॥

The two things which by their existence bring about the distinctness between the world and beyond, which are respectively the condition of motion and rest, which are distinct in function, but same or identical in nature and *pradeśa*, are *dharma* and *adharma*. These are uncreated and of the same magnitude as *lokākāśa*.

Commentary by A. Chakravarti

The existence of these two principles must be postulated as the necessary conditions of the world for without this there will be neither motion nor rest among things. There will be neither the world nor beyond. If the material particles and *jīvas* are not kept together as a system, then they will get scattered through the whole place resulting in sheer chaos. There will be no definite world. There will be neither the beyond (*āloka*). The difference between *loka* and *āloka* is entirely due to the coherent system of molecules and *jīvas*, conditioned by these principles. *Dharma* and *adharma* are said to be distinct because of the difference in function. The former is the condition of motion, the latter of rest. But they are quite similar in nature and are indistinguishable because of their non-exclusive coexistence in space. They are in themselves *niṣkriya-dravyas*. Non-active and non-functional and yet condition the things living and non-living in their motion and rest. For this reason, they are limited entirely to the world. Their function will not be felt beyond the world for the simple reason that there are no things beyond.

88

एयरसवण्णगंधं दो फासं सद्दकारणमसद्दं ।
खंधंतरिदं दव्वं परमाणुं तं वियाणेहि ॥

eyarasavaṇṇagaṁdhaṁ do phāsaṁ saddakāraṇamasaddaṁ ।
khaṁdhaṁtaridaṁ davvaṁ paramāṇuṁ taṁ viyāṇehi ॥

Dharmāstikāya does neither move itself nor effect motion in other substances. But it becomes the causal condition in the movement or motion of *jīvas* (living beings and *pudgalānām* (material objects)) moving on their own.

Commentary by Amṛtacandra

This verse describes that even though *dharma* and *adharma* are the causal conditions (*hetu*) in motion and rest, yet they in themselves are extremely or excessively apathetic, indifferent, inert and passive (*udāsīna*).

Commentary by A. Chakravarti

It is determined that *dharma* and *adharma* being in themselves entirely neutral from the external condition of motion and rest respectively. *Dharmāstikāya* itself is incapable of movement nor can it be a substantive cause of motion in other things such as physical objects and living beings. In what way can it be said to condition motion? Certainly not like the horses that, while themselves running, indirectly cause the motions of persons on their back. *Dharmāstikāya* does not carry things; locomotion of things and persons is not brought about by *dharma* for it is a *niṣkriya dravya*. But it behaves like water which by its mere presence is the condition of motion in fishes, i.e. *dharmāstikāya* by its mere existence conditions motion without being the substantive cause of motion. *Jīvas* and *pudgalas* have motion because of their own substantive causes and yet motion in them will be impossible, but for the eternal condition of *dharmāstikāya*.

Similarly, *adharma* being neutral in itself is the external condition of rest. It is merely an external causal condition (*bahiraṅga-hetu*). Just as the earth is the standing place for horses and the shadow of a tree, the place of rest for the pilgrims, so *adharma* is the non-substantive supportive factor (*nimitta*), auxiliary cause (*nimitta kāraṇa*) or external condition of things at rest.

89

उवभोज्जमिंदिएहिं य इंदिय काया मणो य कम्माणि ।
जं हवदि मुत्तमण्णं तं सव्वं पुग्गलं जाणे ॥

uvabhojjamiṁdiehiṁ ya iṁdiya kāyā maṇo ya kammāṇi ।
jaṁ havadi muttamaṇṇaṁ taṁ savvaṁ puggalaṁ jāṇe ॥

To whatever things there is motion, to the very same things there can be rest, or cessation of motion [and to whatever things there is] rest, to the very same things there can be motion], such things of their own states-of-evolution (*svaka-pariṇāma*) effect their motion or rest.

Commentary by Amr̥tacandra

The verse explains that *dharma* and *adharma* are indifferent, inert, passive or apathetic (*udāsīna*) condition (*hetu*) of motion and rest respectively. In reality, if *dharma* and *adharma* were the primary condition (*mukhya hetu*) in the motion or rest of *jīva* and *pudgala* then the things that are in motion must continue to be in motion forever. But things do not behave in that way. Things that are at rest must continue to be at rest forever. But things do not behave in that way. Therefore, it is inferred that they (*dharma* and *adharma*) are not the *mukhya hetu* in motion and rest of things, but from external, other-referential point of view (*vyavahāra naya*), they are the *udāsīna* condition (*hetu*).

To the question as to how things effect motion and rest, it is stated that in reality from the internal, self-referential point of view (*niścaya naya*), all such things effect their motion and interest due to their own states-of-evolution (*pariṇāma*) alone.

CHAPTER FOUR

Ākāśa (Space)

90

धम्मत्थिकायमरसं अवण्णगंधं असद्दमप्फासं ।
लोगोगाढं पुट्ठं पिहुलमसंखादियपदेसं ॥

dhammatthikāyamarasaṁ avaṇṇagaṅdhaṁ asaddam-aphāsaṁ ।
logogāḍhaṁ puṭṭhaṁ pihulamasaṅkhādiyapadesaṁ ॥

Now expounding space (*ākāśa*) *dravya*, Kundakunda states: "That which contains or accommodates completely all *jīvas* and *pudgalas* and the remaining substances (*dravyas*) is *loka-ākāśa* (the world space)."

Commentary by Amr̥tacandra

That which is the supportive factor (*nimitta*) in completely accommodating or giving space (*avakāśa*) to all the substances in this *loka*, i.e. universe of six substances, is *ākāśa*, which is of the nature of uncontaminated pure region, area or locality (*viśuddha kṣetra rūpa*).

Space is considered to be an objectively real substance in the Jaina system. It is not Ether, as George Thibaut points out.

91

अगुरुलघुगेहिं सया तेहिं अणंतेहिं परिणदं णिच्चं।
गदिकिरियाजुत्ताणं कारणभूदं सयमकज्जं॥

agurulaghugehiṁ sayā tehiṁ aṇaṅtehiṁ pariṇadaṁ ṇiccaṁ ।
gadikiriyājuttāṇaṁ kāraṇabhūdaṁ sayamakajjaṁ ॥

Jīva, matter (*pudgala*), the principle of motion and that of rest and also time, these are identical or inalienable (*ananya*) not distinct from the universe (*loka*). But that which is the same as the universe, and due to being infinite that *ākāśa* (space), is *ananya* and also other (*anya*) from the *loka* (universe).

Commentary by Amṛtacandra

Jīva, etc. all the substances, except *ākāśa*, are *ananya*, i.e. inalienable with *loka* due to being of limited or restricted (*maryādita*) states-of-evolution (*pariṇāma*), but *ākāśa* (space) due to being infinite is both *ananya* from *loka* and also *anya* (other).

Commentary by A. Chakravarti

The constituent elements of the world are the infinite number of *jīvas,* and the infinite physical objects, the principles of motion and rest, in space and time. Of these, space contains the other five. The space which is coextensive with these objects called *lokākāśa*. But that is only a part of the real space. Beyond the *loka*, there is *ālokākāśa* (infinite space) (*anantākāśa*). This *anantākāśa* is your space. There are no objects animator inanimate in this infinite Beyond. Neither a tiny molecule of matter nor a single *jīva*, would stem beyond the limits of *loka*. The system of objects is held together by the principles of *dharma* and *adharma*. And these principles are confined to *lokākāśa*.

Thus, we have to note that mathematically pure space is

recognized to be possible and real by Jaina thinkers. Arguments against pure space have already been said to be generally fallacious.

92

उदयं जह मच्छाणं गमणाणुग्गहयरं हवदि लोए ॥
तह जीवपुग्गलाणं धम्मं दव्वं वियाणेहि ॥

udayaṁ jaha macchāṇaṁ gamaṇāṇuggahayaraṁ havadi loe ।
taha jīvapuggalāṇaṁ dhammaṁ davvaṁ viyāṇehi ॥

If space in addition to accommodating other things is the causal condition of their motion and rest, then why do the *siddhas* have the tendency to go upwards and come to stay at the summit of the world?

93

जह हवदि धम्मदव्वं तह तं जाणेह दव्वमधमक्खं ।
ठिदिकिरियाजुत्ताणं कारणभदं तु पुढवीव ॥

jaha havadi dhammadavvaṁ taha taṁ jāṇeha davva-madhamakkhaṁ ।
ṭhidikiriyājuttāṇaṁ kāraṇabhadaṁ tu puḍhavīva ॥

The liberated souls, the *siddhas*, stay in the summit of the world – so it is revealed by the great Omniscient Lords (Jinas). Therefore, there cannot be in space either the condition of motion or of rest.

Commentary by Amṛtacandra

Because the liberated souls, the *siddhas* reside at the summit of the world, hence it is decided that the space is not the condition (*hetu*) of motion and rest. The principles of *dharma* and *adharma*, which are the cause of division between *loka* and *āloka*, are the condition (*hetu*) of motion and rest.

94

जादो अलोगलोगो जेसिं सब्भावदो य गमणठिदी।
दो वि य मया विभत्ता अविभत्ता लोयमेत्ता य॥

jādo alogalogo jesiṁ sabbhāvado ya gamaṇaṭhidī ।
do vi ya mayā vibhattā avibhattā loyamettā ya ॥

If space be the condition of motion and rest of [*jīva* and matter (*pudgala*)], then *āloka* would disappear and the growth or increase beyond the limit of *loka*, i.e. the issue of destruction and dissipation of *loka*, the universe or the world.

95

ण य गच्छदि धम्मत्थी गमणं ण करेदि अण्णदवियस्स॥
हवदि गती स प्पसरो जीवाणं पुग्गलाणं च॥

ṇa ya gacchadi dhammatthī gamaṇaṁ ṇa karedi aṇṇadaviyassa ।
havadi gatī sa ppasaro jīvāṇaṁ puggalānaṁ ca ॥

Therefore, *dharma* and *adharma* are the condition of motion and rest respectively, and not of *ākāśa* (space). So is the nature of the universe (*loka*) as revealed by the great Omniscient Lords (Jinas) to their audience (in *samavasaraṇa*), the hall of audience of the *tīrthaṅkaras*.

Commentary by Amr̥tacandra

This is the conclusion of the destruction of the view that *ākāśa*, (space) is the condition of motion and rest. *Dharma* and *adharma* are the causal condition of motion and rest, not of *ākāśa* (space).

96

विज्जदि जेसिं गमणं ठाणं पुण तेसिमेव संभवदि।
ते सगपरणामेहिं दु गमणं ठाणं च कुव्वंति॥

vijjadi jesiṁ gamaṇaṁ ṭhāṇaṁ puṇa tesimeva saṁbhavadi ।
te sagaparaṇāmehiṁ du gamaṇaṁ ṭhāṇaṁ ca kuvvaṅti ॥

Dharma, *adharma* and space are of the same size and dimensions and constitute an inseparable unity. But due to obtaining a separate or different function, they attain unity or oneness (*ekatva*) as also otherness (*anyatva*).

Commentary by Amṛtacandra

Here in this verse, with respect to *dharma*, *adharma* and accommodation of *lokākāśa*, even due to having *ekatva*, they are said to exhibit otherness (*anyatva*) from the nature of things.

Due to having the same size and dimension, and with respect to occupying the same locality (*eka kṣetrāvagaha*), they have *ekatva*; yet (1) they have fundamentally different functions, *dharma* being the condition of motion, *adharma* being the condition of rest and space being the condition of accommodating or giving space (each obtaining separate particular (*viśeṣa*) function) from the external, other-referential point of view (*vyavahāra naya*); and (2) from the internal, self-referential (*niścaya*) point of view, due to obtaining separate *viśeṣa* of the nature of diversity of function and nature, and distinctness of *pradeśa*, there have only otherness (*anyatva*) alone.

Resume or Concluding Remarks

97

सव्वेसिं जीवाणं सेसाणं तह य पुग्गलाणं च ।
ज देदि विवरमखिलं तं लोए हवदि आयासं ॥

savvesiṁ jīvāṇaṁ sesāṇaṁ taha ya puggalāṇaṁ ca ।
ja dedi vivaramakhilaṁ taṁ loe havadi āyāsaṁ ॥

Space, time, *jīva*, *dharma* and *adharma*: these are non-corporeal (*amūrta*) *dravyas* (substances). Matter (*pudgala*) alone is corporeal (*mūrta*) [and has sense qualities]. Again, of these, *jīva* alone is indeed conscious.

Commentary by Amr̥tacandra

Mūrta (corporeal) *dravya* is one which has sense qualities, of touch, taste, smell and colour, and that which is devoid of these qualities is *amūrta* (non-corporeal). That which has the nature of consciousness is *cetana* (conscious being) and that which is devoid of consciousness is *acetana* (non-conscious).

Thus, *ākāśa* is *amūrta*, time is *amūrta*, *jīva* is *amūrta* by nature, but with respect to joining together with *mūrta dravya*, it is also *mūrta*, *dharma* is *amūrta*, *adharma* is *amūrta*, only *pudgala* is *mūrta*. *Ākāśa* is *acetana*, time is *acetana*, *dharma* is *acetana*, *adharma* is *acetana*, *pudgala* is *acetana*, only *jīva* is *cetana*.

98

जीवा पुग्गलकाया धम्माधम्मा य लोगदोणण्णा ।
तत्तो अणण्णमण्णं आयासं अंतवदिरित्तं ॥

jīvā puggalakāyā dhammadhammā ya logadoṇaṇṇā ।
tatto aṇaṇṇamaṇṇaṁ āyāsaṁ aṅtavadirittaṁ ॥

Jīvas and *pudgala*, i.e. physical objects become active (*sakriya*), i.e. causal agents were determined by certain external causal conditions. The remaining *dravyas* (substances) are not causally active. *Jīva* is active because of *pudgala* being the extrinsic causal condition. The *skandhas* (physical bodies) are active due to time being of the external causal condition.

Commentary by Amr̥tacandra

The condition of obtaining differentiation of place or location is that action (*kriyā*) of modification (*paryāya*) of the nature of *parispanda* (evolutions into action). With external causal condition, the *jīva* is active and with external causal condition, *pudgala* is active. *Ākāśa* (space) is inactive (*niṣkriya*), *dharma* is *niṣkriya*, *adharma* is *niṣkriya* and time (*kāla*) is *niṣkriya*.

The external causal condition of *jīvas* being active is the accumulation of combination of *karma* and no-*karma* (physical body). Therefore, *jīva* is active due to *karmas* and *pudgala*. In their absence, the liberated souls, the *siddhas*, are *niṣkriya*. The external causal condition of *pudgala* being active is time which accomplishes or completes the states-of-evolution (*pariṇāma*).

Like *karmas*, time does not have absence at any time; therefore, like the *siddhas*, there does not occur a state of *niṣkriya* with respect to *pudgalas*.

99

आगासं अवगासं गमणट्ठिदिकारणेहिं देदि जदि ।
उढ्ढंगदिप्पधाणा सिद्धा चिट्ठंति किध तत्थ ॥

āgāsaṁ avagāsaṁ gamaṇaṭṭhidikāraṇehiṁ dedi jadi ।
uḍhḍhaṅgadippadhānā siddhā ciṭṭhaṁti kidha tattha ॥

All those objects that are apprehended or perceived by the senses are corporeal (*mūrta*) objects. The remaining objects are *amūrta* (non-corporeal). The mind apprehends both *mūrta* and *amūrta* objects.

Commentary by Amr̥tacandra

In this universe, the soul (*jīva*) apprehends objects of sense qualities of touch, taste, smell, colour and sound, they are at some time (*kadācita*) gross *skandhas*, *kadācita* they are subtle and *kadācita* they are in the form of *paramāṇu*. Thus, due to obtaining different forms, they are perceptible to the senses or not perceptible to the senses. However, due to their suitability or capacity (*yogyatā*) of the perceptible to the senses, they (the objects) always have the good disposition (*sadbhāva*) of being perceptible to senses, are called *mūrta*.

All those remaining objects, whose nature is devoid of touch, taste, smell, etc. due to the absence of having the suitability of capacity, are called *amūrta*.

Both *mūrta* and *amūrta* objects mentioned above, due to having the *sadbhāva* of the *yogyatā* are perceptible to *citta* (mind). The mind, not having fixed or prescribed subject or matter, being *aprāpyakārī* (not having contact with its object) and being the means of sensitive knowledge acquired by means of five senses and mind, and scriptural knowledge, grasps or comprehends both *mūrta* and *amūrta*.

100

जह्मा उवरिट्ठाणं सिद्धाणं जिणवरेहिं पण्णत्तं ।
तह्मा गमणट्ठाणं आयासे जाण णत्थित्ति ॥

jahmā uvariṭṭhāṇaṁ siddhāṇaṁ jiṇavarehiṁ paṇṇattaṁ ।
tahmā gamaṇaṭṭhāṇaṁ āyāse jāṇa ṇatthitti ॥

Relative time (*vyavahāra kāla*) is determined by the changes or states-of-evolution of *jīva* and *pudgala*. These changes or states-of-evolution are the effects of Absolute Time (*dravya kāla*). The former time is ephemeral [having beginning and end] and the latter is eternal or permanent (*nitya*); such are the nature or the characteristics of the two.

Commentary by Amr̥tacandra

This verse describes the nature of relative time (*vyavahāra kāla*) and absolute time (*niścaya kāla*).

Time gradually in successive order modifications from *samaya* (moment) onwards is a relative time; its base, i.e. *dravya*, is absolute time.

The relative time, being a mode of *niścaya kāla*, is measured by changes or states-of-evolution (*pariṇāma*) of *jīva* and *pudgala* being produced by *pariṇāma* of *jīva* and *pudgala*. And *jīva* and *pudgala*, therefore, are called as being external supportive factors (*nimitta bhūtas*) and due to being produced by the good disposition of *dravya kāla* (absolute time) are called "those produced by effects of *dravya kāla*". The purport of saying this is that *vyavahāra kāla* is determined by the *pariṇāma* of *jīva* and *pudgala* and *niścaya kāla*, not being produced by contrary *pariṇāma* of *jīva* and *pudgala* (in other words, the *pariṇāma* of *jīva* and *pudgala* cannot be formed by any other manner or way, therefore) it is so determined.

There, *vyavahāra kāla* is ephemeral because it is only a subtle mode (as much as only moment), (and) *niścaya kāla* is permanent (*nitya*), because it is always indestructible due to being the base of its own qualities and modes.

101

जदि हवदि गमणहेदू आगासं ठाणकारणं तेसिं।
पसजदि अलोगहाणी लोगस्स य अंतपरिवुड्ठी॥

jadi havadi gamaṇahedū āgāsaṁ ṭhāṇakāraṇaṁ tesiṁ ।
pasajadi alogahāṇī logassa ya aṁtaparivuḍṭhī ॥

The designation or appellation of time denotes an existential fact; therefore, *niścaya kāla* is permanent (*nitya*). The other *vyavahāra kāla* originates and decays, and hence due to the flow may also be of long duration.

Commentary by Amr̥tacandra

This verse describes *nitya kāla* and ephemeral (*kṣaṇika*) *kāla*.

This is time, this is time, this or a similar kind of appellation, this kind of (direction or designation and description) is *niścaya kāla*. In reality, displaying as evident its good disposition is permanent (*nitya*). And that which originates and is destroyed together in a single moment is *vyavahāra kāla*. In reality, it is a mode (*paryāya*) of the appellation of "time" of that very particular substance (*dravya viśeṣa*). Even though it is ephemeral, it displays its flow or course and, therefore, from the strength of the [external, other-referential (*vyavahāra*)] point of view may be said to have a long duration, there is no fault in saying that. Describing it as a measure of the whole (*avalika*), a moment or an instant (*palyopama* and *sāgaropama*), etc. from *vyavahāra*, is not denied.

In this way, here it is said that *niścaya kāla*, due to being the nature of *dravya*, is *nitya* and *vyavahāra kāla*, being the nature of mode (*paryāya*), is ephemeral.

102

तह्मा धम्माधम्मा गमणट्ठिदिकारणाणि णागासं ।
इदि जिणवरेहिं भणिदं लोगसहावं सुणंताणं ॥

tahmā dhamādhammā gamaṇaṭṭhidikāraṇāṇi ṇāgāsaṁ ।
idi jiṇavarehiṁ bhaṇidaṁ logasahāvaṁ suṇaṁtāṇaṁ ॥

Time, space (*ākāśa*), *dharma* and *adharma*, matter (*pudgala*) and *jīva*: these things are called *dravyas* (substances). Of these, time alone is devoid of space-points (*pradeśa*).

Commentary by Amr̥tacandra

Like *jīva*, *pudgala*, *dharma*, *adharma* and *ākāśa*, time also has all the characteristics (such as existence and continuity through origin and decay) and so obtain the appellation or name of *dravya*. In this way, there are six *dravyas*. But like *jīva*, *pudgala*, *dharma*, *adharma* and *ākāśa* have the characteristic of extension, i.e. two or more space-points (*pradeśas*), such as the nature of *astikāya*, in the same way *kālāṇus*, though their number is as much as the *pradeśas* of *lokākāśa*, due to being one space-point (*eka-pradeśopama*), it (time) is not an *astikāya*.

That is the reason that in the exposition primarily of five *astikāyas*, the description of time has not been done. But it is known from the states-of-evolution (*pariṇāma*) of *jīva* and *pudgala* (matter) that due to time having its mode (*paryāya*) and it is inferred that due to not having contrary production of the *pariṇāma* of *jīva* and *pudgala*, and due to the fact that such is that time *dravya*, it (time) has been included in substance (*dravya*).

In *Pravacanasāra* 93.2, Kundakunda refers to *tiryak-pracaya*, that which has persisting collection of crosswise generalities or horizontal generalities, and *ūrdhva-pracaya* (vertical generality) in which the series is mono-dimensional or linear order, as compared to multi-dimensional series.

The five *astikāyas* have *tiryak-pracaya*, but time (*kāla*) has only *ūrdhva-pracaya*. It has no extension either directly or indirectly. Hence, it is not an *astikāya*. Though it is not an *astikāya*, it is distinctly a real entity, which accounts for changes in other things.

103

धम्माधम्मागासा अपुब्धभूदा समाणपरिमाणा ।
पुधगुवलद्धिविसेसा करंति एगत्तमण्णत्तं ॥

dhammādhammāgāsā apubdhabhūdā samāṇaparimāṇā ।
pudhaguvaladdhivisesā karaṁti egattamaṇṇattaṁ ॥

Thus, one who after knowing (and having faith) in the *Pañcāstikāya-saṁgraha* – the essence of divine discourses – gives up attachment and aversion (towards worldly things) and realizes freedom from suffering.

Commentary by Amṛtacandra

Here, after stating the fruit of knowing the treatise *Pañcāstikāyasāra*, Kundakunda concludes the exposition thereon.

In reality, the essence of divine discourses is the *Pañcāstikāya-saṁgraha*. He, who after knowing its meaning in a proper manner and determining the nature of its own *jīva astikāya* as an extremely purification of consciousness and seeing the impurity of attachment and aversion as the cause of kārmic bondage, displays its discriminative illumination (*viveka jyoti*), giving up the transformation of attachment and aversion, that person thereby turns away or is averse to the future kārmic bondage and giving up previous kārmic bondage realizes freedom from all worldly sufferings.

104

आगासकालजीवा धम्माधम्मा य मुत्तिपरिहीणा ।
मुत्तं पुग्गलदव्वं जीवो खलु चेदणो तेसु ॥

āgāsakālajīvā dhammādhammā ya muttiparihīṇā ।
muttaṁ puggaladavvaṁ jīvo khalu cedaṇo tesu ॥

Whenever one knowing this truth (that the Self is the greatest reality) and seeks to reach or realize the same, gets free from deluded view (*darśana mohanīya*) and abstains or refrains from attachment and aversion (*caritra-mohanīya*) and finally whose future and previous kārmic bondage is destroyed, he becomes that.

Commentary by Amr̥tacandra

In this verse, the author describes the order of events that leads to liberation (*mokṣa*).

First, he knows that the Self of the nature of purity of consciousness and endeavours to realize the same. As a result, the destruction of deluded view (*darśana moha*) takes place, thereby dawns upon him the radiance of knowledge (*jñāna jyoti*) of the reality of things. Therefore, he abstains or refrains from attachment and aversion (*caritra-mohanīya karma*). Hence, the destruction of past and future kārmic bondage takes place, and then with the absence of the condition (*hetu*) of the Self being bound again, the radiance of the glorious, divine lordship of the Self shines eternally.

Thus, ends Book One of *Pañcāstikāyasāra*.

BOOK TWO

Nine Padārthas or Tattvas

Introductory Remarks of Amr̥tacandra

After ascribing to the knowledgeable persons the pure *ātma-tattva*, i.e. the principle of *karma-nirapekṣa* (non-relational or independent of *karmas*) the pure nature of the Self in the previous book, now starting from nine categories (*padārthas*), or nine basic principles of life (*tattvas*), the author is going to describe the path (*mārga*) [of liberation], by paying tribute to Lord Mahāvīra, who revealed the path.

105

जीवा पुग्गलकाया सह सक्किरिया हवंति ण य सेसा ।
पुग्गलकरणा जीवा खंधा खलु कालकरणा दु ॥

jīvā puggalakāyā saha sakkiriyā havaṁti ṇa ya sesā ।
puggalakaraṇa jīvā khaṁdhā khalu kālakaraṇā du ॥

Bowing my head in reverence to the Omniscient Lord Mahāvīra, the causal factor in liberation, I describe those two primary principles of *jīva* (conscious being) and the non-conscious entity (*ajīva*), particularly, matter (*pudgala*) and the various derivative principles or categories and also the path of liberation (*mokṣa*).

Commentary

The two primary or fundamental categories are associated with each other from beginningless time; the other seven derivatives or secondary categories, which are the resultant categories due to the interaction of the first two. In spite of the subsidiary nature of the seven categories, they are equally important as the first two inasmuch as their knowledge is quite essential to the process of self-development leading to the liberation of the Self.

These seven categories are: merit or virtue (*puṇya*), sin or vice (*pāpa*), influx (*āśrava*), stoppage (*saṁvara*), shedding (*nirjara*) and bondage (*bandha*) of *karmas* and liberation (*mokṣa*).

106

जे खलु इन्दियगेज्झा विसया जीवेहिं हुंति ते मुत्ता।
सेसं हवदि अमुत्तं चित्तं उभयं समादियदि॥

je khalu indiyagejjhā visayā jīvehiṁ huṅti te muttā ।
sesaṁ havadi amuttaṁ cittaṁ ubhayaṁ samādiyadi ॥

Right or enlightened conduct, united or combined with enlightened world-view or right faith and enlightened knowledge, deprived of attachment and aversion, forms the path of liberation to those faithful *jīvas* who realize knowledge of the Self through the five attainments (*labdhis*).

Commentary by Amṛtacandra

(1) Enlightened or right conduct, combined with the right faith and right knowledge; (2) not any conduct united with a deluded faith in perverted knowledge; (3) devoid of attachment and aversion; (4) not imbued with attachment and aversion, which would lead to bondage; (5) the path of liberation, not the path of bondage; (6) the path available to liberable (*bhavya*) souls, i.e. good-natured ones, and not to *abhavyas* of intrinsic evil disposition; (7) the path available to those who have five attainments (*pañca-labdhi*)

(*Gommaṭasāra*, Jīvakāṇḍa, v. 650) and not to those who do not have those attainments; and (8) path to those who have completely eradicated all the four passions (of anger, pride, deceit and greed), not those who have any such passions. In this way, one has to understand eight kinds of paths.

The true insight or enlightened world-view is not possible when there is intensity of deluded *karmas* and the most intense passions called *anantabandhī* passions. However, the weakening of the view-deluding *karmas* is due to discriminative wisdom as to what is beneficial to the soul, that is partly destructive and partly subsidential (*kṣayopaśama*) attainment (*labdhi* no. 1).

After the initial clarity in outlook, the Self may be hankering for good, virtuous or auspicious acts (*puṇya*) and the purity of purification (*viśuddhi*) of psychic disposition is *viśuddhi* (*labdhi* no. 2).

Then, that *jīva* is disposed to seek instructions in the path from a competent master in order to gain and enhance his knowledge, without doubt and error, that is *deśanā* or *upadeśa* (*labdhi* no. 3).

With the acquisition of this sort of efficiency, *jīva* reduces the duration and intensity of various *karmas*. He, thus, advances on the path of purification in the declining stage of *karmas*, this is *prayogya* (*labdhi* no. 4).

The fifth attainment of *karaṇa labdhi* is classified into three stages:

a. *Adhapravr̥ttikaraṇa*: In this process, the Self reduces the duration and intensity of *karmas* to a considerable extent;

b. *apūrva karaṇa*: In this process, the soul attains new insights in regard to the reality of things as it has never experienced before; and

c. *anivr̥ttikaraṇa*: Just after the acquisition of this *labdhi*, there is spiritual awakening and the dawn of enlightened world-view or right belief (*samyak darśana*).

107

कालो परिणामभवो परिणामो दव्वकालसंभूदो ।
दोण्हं एस सहावो कालो खणभंगुरो णियदी ॥

kālo pariṇāmabhavo pariṇāmo davvakālasaṁbhūdo ।
doṇhaṁ esa sahāvo kālo khaṇabhaṅguro ṇiyadī ॥

Belief or conviction in the nine spiritual principles (*tattvas*) or categories (*bhavanam*) is *samyakatva*, i.e. enlightened world-view or right faith, their acquisition is enlightened knowledge, without doubt and error. An attitude of equanimity without attachment and aversion towards the objects of the senses is enlightened/right conduct.

Commentary by Amr̥tacandra

This verse refers to enlightened world-view, knowledge and conduct (called triple jewels).

The enlightened world-view (*samyak darśana*) is the seed of the determination of the pure nature of the Self of the character or nature consciousness. True or enlightened understanding of the nine *tattvas* (categories) on the discontinuation or abstinence of deluded view is enlightened kṇowledge. The equanimous attitude of those mounted on the path particularly in one's own *tattva*, i.e. basic spiritual principle and due to the absence of the impurity of attachment and aversion towards the objects of the senses and mind, is enlightened conduct, which is the seed of the great happiness or bliss (of liberation, *mokṣa*).

In subsequent verses, *niścaya* and *vyavahāra* path of liberation of the three characteristics will be described.[1]

---

[1] Introductory remark of Amr̥tacandra.

108

कालोत्ति य ववदेसो सब्भावपरूवगो हवदि णिच्चो।
उप्पण्णप्पद्धंसी अवरो दीहंतरट्ठाई॥

kālotti ya vavadeso sabbhāvaparūvago havadi ṇicco ।
uppaṇṇappaddhaṁsi avaro dīhaṁtaraṭṭhāī ॥

Jīva and *ajīva* are the two primary *tattvas* or *padārthas*. Then there are seven derivative native *tattvas* (categories), viz. righteousness or merit (*puṇya*), sin (*pāpa*), the inflow (*āśrava*), stoppage (*saṁvara*), shedding or partial association (*nirjara*), bondage (*bandha*) of *karmas* and salvation or liberation (*mokṣa*). These are the nine *tattvas* (categories) (*padārthas*).

Commentary by Amṛtacandra

This verse describes the names of the nine *tattvas* or *padārthas* and their nature.

Of these, *jīva astikāya* alone has the characteristic of consciousness. That which has a characteristic of the absence of consciousness is *ajīva* (non-conscious entity), which is of five kinds. These are: *pudgala* (matter) *astikāya*, the principle of motion (*dharma astikāya*), the principle of rest (*adharma astikāya*), space (*ākāśa astikāya*) and time *dravya*. Due to being brought about and accomplished through separate existence and having different natures, such are two primary or root *tattvas* or *padārthas*.

The seven other *tattvas* or *padārthas* are produced due to the states-of-evolution (*pariṇāma*) of the joining together of *jīva* and *pudgala*. The auspicious (*śubha*) *pariṇāma* of *jīva* is *puṇya* (virtue or merit) and whose *śubha pariṇāma* is the supportive factor (*nimitta*). These *karma pariṇāmas* of matter (*pudgalas*) of the nature of *śubha* acts (*karma*) is *puṇya*. The inauspicious (*aśubha*) *pariṇāma* of *jīva* is sin (*pāpa*) and whose *aśubha pariṇāma* is the supportive factor (*nimitta*), those *karma pariṇāmas* of *pudgalas* is *pāpa*.

The *pariṇāma* of infatuation (*moha*), attachment (*rāga*) and aversion (*dveṣa*) of *jīva*, i.e. inflow (*āśrava*) of *karmas*, and these *moha–rāga–dveṣa pariṇāma* are the *nimitta* of whose, such entry or admission of *karma pariṇāmas* of *pudgalas*, through the activities of mind, speech and body (*yoga*) are *āśrava*. The stopping of the soul's *moha–rāga–dveṣa pariṇāma* is *saṁvara* and the stopping of that *moha–rāga–dveṣa pariṇāma* is the *nimitta* of whose, such stopping of *karma pariṇāma* of material particles through *yoga* is *saṁvara*.

That which is capable of destroying the strength and vigour of *karmas*, such twelve kinds of austerities, both external and internal, are the cause of the growth of pure conscious attentiveness (*śuddha upayoga*), that is shedding (*nirjara*) of *karmas*. Their influence (i.e. that growth of *śuddha upayoga* is *nimitta*), *nirjara* takes place because of such partial dissociation of *karma pudgalas*.

Due to the affectionate (*snigdha*) *moha–rāga–dveṣa pariṇāma* of *jīva* that attracts bondage (*bandha*) of *karmas* and due to those being the *nimitta* of the transformation of *pudgalas* and their deep immersion with *jīva*, as one another, that is *bandha*. The attainment of the excessive purity of the Self, that is liberation (*mokṣa*) and the excessive separation of *karma pudgalas* from *jīva*, that is *mokṣa*.

109

एदे कालागासा धम्माधम्मा य पुग्गला जीवा ।
लब्भंति दव्वसण्णं कालस्स दु णत्थि कायत्तं ॥

ede kālāgāsā dhammādhammā ya puggalā jīvā
labbhaṁti davvasaṇṇaṁ kālassa du ṇatthi kāyattaṁ ॥

Jīvas are of two kinds: mundane (*saṁsārī*), i.e. the impure souls and liberated, i.e. pure souls (*TS* II.10). In reality, both of them have the nature of consciousness and are characterized by conscious attentiveness or the manifestation or function of consciousness (*upayoga*), the mundane souls with the use of the body and the liberated souls without the body.

Commentary by A. Chakravarti

In this verse, Kundakunda emphasizes the fact that time has *astitva* (existence), but not the *kāyatva* speciality. It is one of the six *dravyas*. The name *dravya* is applicable to all those existences which have continuity through origin and decay. Time has such a nature. Therefore, it shares the name *dravya* with the other things herein enumerated. But the other *dravyas* are also called *astikāyas* because they have an extension (*bahupradeśa*). But time has no such *bahupradeśa*. Therefore, it is not an *astikāya*.

This nature of time is described in mathematical terms by Kundakunda in *Pravacanasāra*. Time has *ūrdhva pracaya*, whereas the other *dravyas* have *tiryak pracaya*. In the language of modern mathematics, the former term corresponds to mono-dimensional assymetrical series; the latter to a multi-dimensional series. According to mathematical philosophy, an extension is a multi-dimensional series. *Kāyatva*, therefore, is generated by *tiryak pracaya*, but *ūrdhva pracaya* cannot be associated with an extension since it is unilateral. Therefore, time is not an *astikāya*.

110

एवं पवयणसारं पंचत्थियसंगहं वियाणित्ता ।
जो मुयदि रागदोसे सो गाहदि दुक्खपरिमोक्खं ॥

evaṁ pavayaṇasāraṁ paṅcatthiyasaṁgahaṁ viyāṇīttā ।
jo muyadi rāgadose so gāhadi dukkhaparimokkhaṁ ॥

The five kinds of immobile *jīvas* are one-sensed *jīvas*. These are earth-bodied, water-bodied, fire-bodied, wind-bodied and plants. These are many in number. They yield to their respective *jīvas* only one kind of feeling, touch; and that too associated with excessive infatuation.

Commentary by Amṛtacandra

The five kinds of earth-bodied, water-bodied, fire-bodied, wind-bodied and plants, which due to bondage (of *karmas*), are vitalized by *jīvas*. Due to the intermediate differentiations, they are many in number; even then they are all the states-of-evolution of *pudgala* and, having *kṣyayopaśama*, i.e. partly destructive and partly subsidential of touch-sense obscuring, being formed or composed of external touch sense of *jīvas* and due to being primarily of *karma-phala-cetana*, i.e. experiencing merely the consciousness of the fruits of one's past *karmas*, a passive experience of agreeable and disagreeable phenomenon, is excessively imbued with infatuation, make possible only the attainment of touch-sense.

111

मुणिऊण एतदट्ठंतदणुगमणुज्झदो णिहदमोहो।
पसमियरागद्दोसो हवदि हदपरावरो जीवो॥

muṇiūṇa etadaṭṭhaṅtadaṇugamaṇujjhado ṇihadamoho ।
pasamiyarāgaddoso havadi hadaparāvaro jīvo ॥

Three of these five immobile one-sensed *jīvas*, viz. earth-bodied, water-bodied and plant-bodied, after having joining together with immobile bodies and the other two, i.e. fire-bodied and wind-bodied *jīvas*, are mobile (*trasa*); they all are to be known as one-sensed *jīvas*, devoid of mental states-of-evolution.

Commentary by A. Chakravarti

Kundakunda describes the order of events that culminate in *mokṣa*. The greatest thing mentioned in this scripture is the Self. Whoever understands the glorious nature of self-consciousness cannot but seek to realize the same. On account of that effort, the wall round the faculty of faith or *darśana mohanīya* gets removed. By the removal of this ignorance, dawns the knowledge of the Self. Through self-knowledge, the rooting out of desire leads to aversion, the destruction of *caritra mohanīya*. Hence, the shattering of kārmic shackles, leads to freedom, which, in turn, radiates the eternal glory of the Divine Person.

112

अभिवंदिऊण सिरसा अपुणब्भवकारणं महावीरं।
तेसिं पयत्थभंगं मग्गं मोक्खस्स वोच्छामि॥

abhivaṁdiūṇa sirasā apuṇabbhavakāraṇaṁ mahāvīraṁ ।
tesiṁ payatthabhaṅgaṁ maggaṁ mokkhassa vocchāmi ॥

These earthy-bodied, etc. five kinds of *jīvas* are devoid of mental states. They have a single sense of touch. So are they described in the scriptures.

Commentary by Amr̥tacandra

The earth-bodied, etc. *jīvas*, due to the *kṣayopaśama* (see *supra PKS* 110) of touch-sense obscuring and the coming into operation or rise (*udaya*) of the obscuring of the remaining senses and of the rise (*udaya*) of mental obscuring, they are one-sensed, devoid of mind.

113

सम्मत्तणाणजुत्तं चारित्तं रागदोसपरिहीणं।
मोक्खस्स हवदि मग्गो भव्वाणं लद्धबुद्धीणं॥

sammattaṇāṇajuttaṁ cārittaṁ rāgadosaparihīṇaṁ ।
mokkhassa havadi maggo bhavvāṇaṁ laddhabuddhīṇaṁ ॥

Just as life or consciousness is associated with the activity devoid of mental intelligence or thought in the following cases: growing in eggs or the mother's womb and the man in faint (*mūrcchā*), so also the mono-sensic organisms are fainted, unconscious, though having life or consciousness. Thus, be it understood.

Commentary by Amr̥tacandra

This verse is the description of the example of the one-sensed *jīvas* who have existence of consciousness.

114

एवं जिणपण्णत्ते सद्दहमाणस्स भावदो भावे।
पुरिसस्साभिणिबोहे दंसणसद्धो हवदि जुते॥

evaṁ jiṇapaṇṇatte saddahamāṇassa bhāvado bhāve ।
purisassābhiṇibohe daṅsaṇasaddho havadi jute ॥

The sea snail, the cowrie-shell fish, the conch-shell fish, mother of pearl and the earth worm are organisms which have two senses, touch and taste. Hence, they are two-sensed *jīvas*.

Commentary by Amr̥tacandra

Due to the *kṣayopaśama* of the touch and taste obscuring senses and the coming into operation or rise (*udaya*) of the remaining senses obscuring and *udaya* of mental obscuring sea snail, etc. are the two-sensed organisms, devoid of mind so say the knowers of touch and taste senses.

115

सम्मत्तं सद्दहणं भावाणं तेसिमधिगमो णाणं।
चारित्तं समभादो विसयेसु विरूढमग्गाणं ॥

sammattaṁ saddahaṇaṁ bhāvāṇaṁ tesimadhigamo ṇāṇaṁ ।
cārittaṁ samabhādo visayesu virūḍhamaggāṇaṁ ॥

The louse, the bug, the red and ordinary scorpion, the ant and other insects, having three senses of touch, taste and smell, are three-sensed organisms.

Commentary by Amr̥tacandra

Due to the *kṣayopaśama* of the touch, taste and smell-obscuring senses, and the coming into operation or rise (*udaya*) of the remaining senses and the arising of mental obscuring, the louse, etc. are three-sensed *jīvas* having touch, taste and smell; but they are devoid of mind.

116

जीवाजीवा भावा पुण्णं पावं च आसवं तेसिं।
संवरणिज्जरवंधो मोख्खो य हवंति ते अट्ठा॥

jīvājīvā bhāvā puṇṇaṁ pāvaṁ ca āsavaṁ tesi ।
saṁvaraṇijjaravaṅdho mokhkho ya havaṅti te aṭṭhā ॥

Again, the gadfly, the mosquito, the fly, the bee, the beetle and the dragonfly and the butterfly, etc. experience touch, taste, smell and sight, and they are four-sensed *jīvas.*

Commentary by Amr̥tacandra

Due to the *kṣayopaśama* (see *PKS* 110) of touch, taste, smell and sight-obscuring senses and the coming into operation or rise (*udaya*) of sound obscuring and the arising of mental obscuring, the gadfly, etc. are four-sensed having touch, taste, smell and sight. They are *jīvas* devoid of mind.

117

जीवा संसारत्था णिव्वादा चेदणप्पगा दुविहा।
उवओगलक्खणा वि य देहादेहप्पवीचारा॥

jīvā saṁsāratthā ṇivvādā cedaṇappagā duvihā ।
uvaogalakkhaṇā vi ya dehādehappavīcārā ॥

Celestial beings, human beings, denizens of hell and higher animals, all these have five senses: colour, taste, smell, touch and sound. Some of them are water animals, some land animals and some birds of the air. There are very strong ones also among these.

Commentary by Amr̥tacandra

Due to the *kṣayopaśama* of the touch, taste, smell, colour and sound obscuring senses and the coming into operation, or rise (*udaya*) of mental obscuring, some are devoid of mind, and some, due to the *kṣayopaśama* of the mental obscuring, with mind (five-sensed) *jīvas*.

Among these, celestial beings, human beings and denizens of hell are with mind; *tiryañca*, i.e. subhuman beings are of both kinds, viz. with mind and without mind.

The classification of *jīvas* according to sense organs is based more upon the behaviour of insects and animals than on the structure.

118

पुढवीय उदगमगणी वाउवणप्फदिजीवसंसिदा काया।
देंति खलु मोहबहुलं फासं वहुगा वि ते तेसिं॥

puḍhavīya udagamagaṇī vāuvaṇapphadijīvasaṁsidā kāyā ।
deṁti khalu mohabahulaṁ phāsaṁ vahugā vi te tesiṁ ॥

Celestial beings are of four classes. Human beings are of two classes: inhabitants of *karma-bhūmi*, i.e. the place where agriculture, etc. are followed by men for their sustenance, and those of *bhoga-bhūmi*, i.e. the regions where there is enjoyment only, i.e. people do not have to work for their sustenance, and the arts of agriculture, etc. are neither necessary nor known. All that the people want, they get from the wishing-trees called *kalpa-vr̥akṣas*. *Tiryañca*, i.e. sub-human beings (animals), are of numerous kinds. The denizens of hell are of seven kinds according to their respective regions.

Commentary by Amr̥tacandra

Due to the *nimitta* of coming into operation or rise (*udaya*) of the *nāma karma* of celestial condition of existence (*gati*) or the *karma* celestial duration of life (*āyu*), a celestial being is (*deva*) of four kinds. These are: residential (*bhavanavāsī*); (2) peripatetic (*vyantara*); (3) stellar (*jyotiṣka*); and (4) heavenly (*vaimānika*) (*TS* IV.1).

The human beings are of two kinds: (1) those born in *karma*-

bhūmi; and (2) those born in *bhoga-bhūmi*. As for *tiryañca*, subhuman beings, these are of numerous kinds: that of two-sensed snail (*śambūka*), the three-sensed louse (*jun*), and the four-sensed gadfly living on earth, water animals, birds, *parisarpa*, *catuṣpāda*, etc. The denizens of hell are those living in seven hells. These are: *ratna-prabhā*, *śarkarā-prabhā*, *vāluka-prabhā*, *paṅka-prabhā*, *dhūma-prabhā*, *tamaḥ-prabhā* and *mahatmaḥ-prabhā*.

Of these celestial beings, human beings and the denizens of hell have only five senses alone. Of the *tiryañca* (subhuman beings, animals, etc.), some have five senses and some are one-sensed, two-sensed, three-sensed and four-sensed also.

119

एदे जीवणिकाया पंचविहा पुढविकाइयादीया ।
मणपरिणामविरहिदा जीवा एगेंदिया भणिया ॥

ede jīvaṇikāyā pañcavihā puḍhavikāiyadīyā ।
maṇapariṇāmavirahidā jīvā egeṅdiyā bhaṇiyā ॥

When the existing *karmas* determining the *gati* (condition of existence) and the age of a *jīva* decay, then that *jīva* gets into another *gati* with a different duration of life as determined by the thought-paint (*leśyā*).

Commentary by Amṛtacandra

Jīva, whose state of existence is determined by a particular *karma*, known as *gati-nāma karma* and duration of life determined by *āyuṣa karma* gradually decays and these two exhaust themselves, then the thought-paint, which is the activity of *yoga* (the activity of mind, speech and body) conciliated or diverted by anger, passions (*kaṣāya*), is the seed or cause of *jīva* to enter into another *gati* or state/condition of existence with a different duration of life. Therefore, *jīvas* obtain other *gati* and other duration of life geneally determined by *karmas* acquired during the life.

But the fundamental factor which immediately determines the passage into the other stage must be determined by the *gati-nāma karma* and *āyuṣa karma* of the succeeding stage. *Jīva*, before quitting a particular *gati*, acquires these two fresh *karmas* determined by the last thought-paint (*leśyā*). It implies merely the particular activity of *yoga*, or mind, speech and body manifesting in the last desire or aspiration of life. Those who do not experience the real nature of the Self (*ātmā*) wanders in the transmigratory existences of life and deaths.

120

उंडेसु पवड्ढंता गब्भत्था माणुसा य मुच्छगया ।
जारिसया तारिसया जीवा एगेंदिया णेयाः ॥

uṁḍesu pavaḍḍhaṁtā gabbhatthā māṇusā ya mucchagayā ।
jārisayā tārisayā jīvā egeṁdiyā ṇeyāḥ ॥

The aforesaid different classes of *jīva* are again said to be of two kinds: the mundane souls incarnate in the body and those devoid of bodies, i.e. the liberated souls (*siddhas*). The mundane (*saṁsārī*) souls are again of two kinds: *bhavyas*, which have the capacity to get liberation, and those, who do not have such a capacity, are called *abhavyas*.

Commentary by Amṛtacandra

All the mundane souls, those with bodies and those without bodies, such as *siddhas*, glorious, divine Lords (*bhagavanta*) who are pure souls. Even with respect to the one kind of mundane souls, who revolve or use body, are of two kinds: like the green gram (*mūṅga*) which are capable of being cooked or not cooked, those ones, who have the capacity of obtaining the pure Self, are called *bhavya*, and those who are devoid of such a capacity, they are called *abhavya*.

121

संवुक्कमादुवाहा संखा सप्पी अपागदा य किमी ।
जाणंति रसं फासं जे ते वे इंदिया जीवाः ॥

saṁvukkamāduvāhā saṅkhā sappī apāgadā ya kimī ।
jāṇaṁti rasaṁ phāsaṁ je te ve iṁdiyā jīvāḥ ॥

The five senses are not the essence of *jīva*, and the six kinds of bodies (the five earth-bodied, immobile bodies and the mobile body) described in the scriptures are also not the essence of *jīva*. Whatever in the midst of these manifests as consciousness, that is called by the name of *jīva*.

Commentary by Amr̥tacandra

The one-sided earth-bodied, etc. sense organs, due to deep immersion of *jīva* and *pudgala* (matter) with one another since beginningless time, are called *jīva* from the external, other-referential point of view (*vyavahāra naya*). But from the internal, self-referential point of view (*niścaya naya*), touch, etc. senses and earth, etc. bodies, in the absence of the characteristic nature of consciousness, are not *jīvas*. But in the midst of these sense organs and physical conditions, there is the principle of consciousness of the knowing nature, which illumines itself and other objects (*sva-para prakāśaka*) (*Svayambhū Stotra* 63), that in some respect being the identity between the quality (*guṇa*) and that having that quality (*guṇa–guṇī kā abheda*) is described as *jīva*.

122

जूगागुंभीमक्कणपिपीलया विच्छियादिया कीडा ।
जाणंति रसं फासं गंधं ते इंदिया जीवा ॥

jūgāguṁbhīmakkaṇapipīlayā vicchiyādiyā kīḍā ।
jāṇaṁti rasaṁ phāsaṁ gaṁdhaṁ te iṁdiyā jīvā ॥

What knows and perceives the various objects and is desirous of happiness and dreads pain or suffering, acts beneficially and harmfully and experiences the fruits thereof – that is *jīva.*

Commentary by Amr̥tacandra

In this verse, the extraordinary work of *jīva* from others is described.

What is mentioned as consciousness in the previous verse is here (in this verse) indicated by several purposeful activities which have meaning only with reference to consciousness. Thus, *jīva* is the causal agent or doer (*kartā*) of knowledge and perception. The related matter is not the *kartā* of these functions, just as space (*ākāśa*), etc. do not belong to *jīva.*

Due to the onset of definite intention or resolve of the nature of change or modification of consciousness, the soul has the tendency to desire happiness and to avoid pain. *Jīva* alone is the causal agent (*kartā*) of continuing beneficial activity and to discontinue harmful activity, not any other. And *jīva* alone is the *kartā* of the activity of consumption or enjoyment of the pleasant and unpleasant sense objects due to the fruitition of auspicious (*śubha*) and inauspicious (*aśubha*) *karmas* and of the activity of one's own states-of-evolution of the nature of pleasure and pain, not any other thing.

Thus, it has been explained that the soul is considered to do the above-mentioned extraordinary activities, not matter (*pudgala*).

123

उद्दंसमसयमविखयमधुकरभमरा पतंगमादीया ।
रूपं रसं च गन्धं फासं पुण ते वि जाणंति ॥

uddaṁsamasayamavikhayamadhukarabhamarā pataṅga-mādīyā ।
rūpaṁ rasaṁ ca gaṁdhaṁ phāsaṁ puṇa te vi jāṇaṁti ॥

Thus, having seen the nature of *jīva* through the numerous and distinct characteristics and modes, now grasp the nature of non-living (*ajīva*), that is absolutely devoid of knowledge.

Commentary by Amṛtacandra

Thus, from the external, other-referential point of view (*vyavahāra naya*), *jīva* and its modifications are to be studied under different principles or expositions of soul classes (*jīva-sthāna*, 98 soul classes), stages of moral and spiritual development (*guṇasthāna*, 14 kinds) and *mārganā-sthāna* (14 kinds, with subdivisions of 74 kinds). In all these multifarious changes there is the one constant character of *jīva* (consciousness).

From the internal self-referential point of view (*niścaya naya*), at some time impure due to the transformation of consciousness into infatuation, attachment and aversion, and at some time due to the absence of impurity, and being of the nature of purity of consciousness, knows *jīva* (consciousness).

Thus, after knowing *jīva*, know that the absence of consciousness is the mark of *ajīva*, which is at some time joined together with *jīva* and at some time not joined together with *jīva*.

124

सुरणरणारयतिरिया वण्णरसप्फासगंधसद्दण्हू ।
जलचरथलजरखचरा वलिया पंचेंदिया जीवा ॥

suraṇaraṇārayatiriyā vaṇṇarasappāsagaṅdhasaddaṇhū ।
jalacarathalajarakhacarā valiyā pañceṅdiyā jīvā ॥

Space, time, matter, the principle of motion and the principle of rest: these have no life principles in them. They are absolutely without consciousness. Only *jīva* alone has consciousness.

Commentary by Amṛtacandra

The quality, particularly of consciousness is not present in space, time, matter and the principles of motion and rest because in those substances, they have in general non-conscious entities (*ajīva*) and this non-consciousness in general is of space, etc. alone, because only *jīva* has consciousness-in-general.

The above statement is critical of both idealist monism which assigns "unreality" to the objective reality of the world comprising of individual selves and material objects and maintains that consciousness alone is the only reality, as also materialist monism which maintains that matter alone is real and everything else is unreal and derivative or that the conscious states or mental events either do not exist or even if they are acknowledged to exist, they must in some sense be reducible or identical to physical states.

The Jaina world-view, which recognizes that there are two realities in the world, viz. the reality of *jīva* (consciousness) and the reality of non-conscious entities, which are grouped under *ajīva padārthas*, which lack consciousness, is based on realistic considerations and is quite natural and logical.

125

देवा चउण्णिकाया मणुया पुण कम्मभोगभूमीया।
तिरिया बहुप्पयारा णेरइया पुढविमेयगदा॥

devā cauṇṇikāyā maṇuyā puṇa kammabhogabhumīyā ।
tiriyā bahuppayārā ṇeraiyā puḍhavimeyagadā ॥

Wherever such attributes [of life], as awareness of pleasure and pain, effort or exertion for the beneficial and fear of harmful activity, never take place, that the wise ones describe *ajīva* (non-living).

Commentary by Amṛtacandra

In the case of space (*ākāśa*), etc. there is always non-availability of fundamental characteristics of consciousness, such as feeling of pleasure and pain, beneficial activity, and avoidance or fear of harmful activity; therefore, it is determined that due to consciousness in general, not existent, space, etc. are non-conscious entity (*ajīva*).

126-127

खीणे पुण्वणिबद्धे गदिणामे आउसे च ते वि खलु ।
पापुण्णंति य अण्णं गदिमाउस्सं सलेसवसा ॥

khīṇe puṇvaṇibaddhe gadiṇāme āuse ca te vi khalu ।
pāpuṇṇaṁti ya aṇṇaṁ gadimāussaṁ salesavasā ॥

एदे जीवणिकाया देहप्पविचारमस्सिदा भणिदा ।
देहविहूणा सिद्धा भव्वा संसारिणो अभव्वा य ॥

ede jīvaṇikāyā dehappavicāramassidā bhaṇidā ।
dehavihūṇā siddhā bhavvā saṁsāriṇo abhavvā ya ॥

An organization or form, a quantity, mass or structure, colour, taste, touch, smell and sound, these are various qualities and modes (*paryāyas*), these are all produced or brought about by the matter (*pudgala*) substance.

Understand that as *jīva*, which has no sense qualities of taste, colour, smell, touch and sound, has no definite bodily shape or form either, and cannot be apprehended by a symbol or sense organ and is endowed with consciousness.

Commentary by Amr̥tacandra

The Self with the joining together of the body is (1) in reality due to having attributes of touch, taste, smell, colour and sound, form and structure, which can be apprehended through the senses, that is *pudgala* (matter), *dravya* (substance); and (2) that which due to being devoid of the sense qualities of touch, taste, smell, colour, and being without any definite shape or form, and being beyond perception through the senses and because of the characteristic of consciousness, and that which is distinct from the non-conscious entities (*jīva*) with form or formless, that is *jīva dravya*.

In this way, the real distinction between *jīva* and *ajīva* is expounded as a condition of the well-known path for those having enlightened or right knowledge.

128-130

ण हि इंदियाणि जीवा काया पुण छप्पयार पण्णत्ता ।
जं हवदि तेसु णाणं जीवोत्ति य तं परूवंति ॥

ṇa hi indiyāṇi jīvā kāyā puṇa chappayāra paṇṇattā ।
jaṁ havadi tesu ṇāṇaṁ jīvotti ya taṁ parūvaṅti ॥

जाणदि पस्सदि सव्वं इच्छदि सुक्खं विभेदि दुक्खादो ।
कुव्वदि हिदमहिदं वा भुंजदि जीवो फलं तेसिं ॥

jāṇadi passadi savvaṁ icchadi sukkhaṁ vibhedi dukkhādo ।
kuvvadi hidamahidaṁ vā bhuṅjadi jīvo phalaṁ tesiṁ ॥

एवमभिगम्म जीवं अण्णेहिं वि पज्जएहिं बहुगेहिं ।
अभिगच्छदु अज्जीवं णाणंतरिदेहिं लिंगेहिं ॥

evamabhigamma jīvaṁ aṇṇehiṁ vi pajjaehiṁ bahugehiṁ ।
abhigacchadu ajjīvaṁ ṇāṇaṅtaridehiṁ liṅgehiṁ ॥

The mundane soul indeed has it states-of-evolution and naturally it has attachment and aversion. On account of the psychic dispositions, the kārmic matter clings to *jīva*. The kārmic bondage leads to the human, etc. conditions of existence (*gati*) and the soul then wanders in the transmigratory cycle of life and deaths.

Entering into the *gati*, *jīva* has an appropriate body; being embodied, he gets the senses. Through the senses, the objects of the senses are pursued. From the perception of pleasant and unpleasant sense objects, one acquires attachment and aversion.

From such mental states of attachment and aversion, the transmigratory cycle begins again. [Thus, attachment brings *karma*; *karma* leads to *gati*; *gati* means body; body implies senses; senses lead to perception; and perception of sense objects again leads to attachment and aversion and so on *ad infinitum*.]

But this transmigratory cycle ends in the case of some *jīva*, whereas it is unending in the case of others. But it has no beginning in either case, i.e. in the former case it is beginningless with, and in the latter case, it is beginningless and endless. Thus, it is spoken by the Omniscient Lord (Jina).

Commentary by Amr̥tacandra

In this *loka*, due to the beginningless *upādhi* of kārmic bondage, the mundane souls experience the states-of-evolution (*pariṇāma*) of attachment and aversion. These psychic states, in turn, bring about the acquisition of material *karma*; *karma* leads to hellish, etc. *gati*, i.e. condition of worldly existence. *Gati* means embodied existence. Body implies senses, which leads to perception of pleasant and unpleasant sense objects. This perception generates the affective states of attachment, etc. In this way, the cause and effect of one another series of *jīva pariṇāma* and kārmic *pudgala* (matter) *pariṇāma*, the *karma* net (*karma-jala*) of transmigratory cycle continues again and again for some *jīva*. It is without beginning and end (*anādi-ananta*) and in case of some other *jīva*, it is beginningless but having an end (*anādi-anta*).

Thus, *pudgala pariṇāma* acts as a supportive factor (*nimitta*) *pariṇāma*. Now the author describes merit (*puṇya*), etc. seven *padārthas* of *jīva pariṇāma* and *jīva-pariṇāma* acts as *nimitta* of *pudgala*.

CHAPTER FIVE

Puṇya and Pāpa (Virtue/Merit and Vice)

131

आगासकालपुग्गलधम्माधम्मेसु णत्थि जीवगुणा ।
तेसिं अचेदणत्तं भणिदं जीवस्स चेदणदा ॥

āgāsakālapuggaladhammādhammesu ṇatthi jīvaguṇā ।
tesiṁ acedaṇattaṁ bhaṇidaṁ jīvassa cedaṇadā ॥

Whatever *jīva* has – infatuation (*moha*), attachment and aversion, or the contentment of thought (*citta prasannatā*) – has the state-of-evolution (*pariṇāma*) of auspicious (*śubha*) or inauspicious (*aśubha*).

Commentary by Amr̥tacandra

This verse describes the nature suitable for merit or virtue (*puṇya*) and sin (*pāpa*).

The impure *pariṇāma* brought about by the maturity or fruition of view deluding (*darśana mohanīya*) is infatuation (*moha*) [which clouds the faculty of perception and belief]. The many kinds of conduct-deluding (*caritra-mohanīya*) affection/liking and lack of affection/liking, leads to attachment (*rāga*) and aversion (*dveṣa*). The purification of *pariṇāma* due to the coming into operation or rise (*udaya*) of mild *caritra-mohanīya*, that is the *pariṇāma* of contentment of thought (the *pariṇāma* of the mind being unstained or unsullied).

In this way, he who has *moha, rāga, dveṣa* (contentment of mind – *citta-prasāda pariṇāma*), he certainly has *śubha* or *aśubha pariṇāma.* Among these, where there is excellent attachment (*praśasta rāga*) and *citta-prasāda*, there is *śubha pariṇāma*; and where there is *moha, dveṣa* and absence or a lack of excellent attachment, there is *aśubha pariṇāma.*

132

सुहदुक्खजाणणा वा हिदपरियम्मं च अहिदभीरुत्तं।
जस्स ण विज्जदि णिच्चं तं समणा विंति अज्जीवं॥

suhadukkhajāṇaṇā vā hidapariyammaṁ ca ahidabhīruttaṁ ।
jassa ṇa vijjadi ṇīccaṁ taṁ samaṇā viṅti ajjīvaṁ ॥

The *śubha pariṇāma* of *jīva* is virtuous or merit (*puṇya*) and *aśubha pariṇāma* is sin or vice (*pāpa*). Due to these two mental states of *jīva*, the *kārmic* molecules undergo modifications as physical *karma* (*dravya karma*).

Commentary by Amṛtacandra

Pure thought is psychic *puṇya* (*bhāva puṇya*), which is the condition of physical (*dravya*) *puṇya*, the kārmic matter facilitating the *śubha pariṇāma* of *jīva*, the causal agent or doer. In the same way, evil thought is psychic (*bhāva*) *pāpa*. This conditions physical (*dravya*) *pāpa*, which aggravates the evil dispositions of the soul.

Thus, the particular class of *pariṇāma* of material (*pudgala*) *karma* (the pleasure feeling *karma*) class, etc. of primary *pariṇāma*), in which the *śubha pariṇāma* of *jīva* is the supportive factor (*nimitta*), they are *dravya puṇya*. The particular class of *pariṇāma* of material *karma* (the pain-feeling *karma*), in which the *aśubha pariṇāma* of *jīva* is the *nimitta*, is *dravya pāpa*.

In this way, both corporeal or physical (*mūrta*) and non-corporeal or pscychic (*amūrta*) conditions of the Self (*ātmā*) are shown, that is described from external, other-referential (*vyavahāra*) and internal, self-referential (*niścaya*) points of view. Thus, *bhāva karma* (psychic *karma*) and *dravya karma* (physical *karma*) are mutually interacting as the cause and effect, each of the other.

133

संठाणा संघादा वण्णरसप्फासगंधसद्दा य।
पोग्गलदव्वप्पभवा होंति गुणा पज्जया य बहू॥

saṁṭhāṇā saṅghādā vaṇṇarasapphāsagaṁdhasaddā ya ।
poggaladavvappabhavā hoṁti guṇā pajjayā ya bahū ॥

Because the fruits of *karma*, the objects of happiness or misery are experienced by *jīva* through the sensations like touch, the *karmas* are physical and *mūrta*.

Commentary by Amṛtacandra

The fruits of *karma*, which are the conditions (*hetu*) of happiness or misery, experienced by *jīva* through the sense organs, are also physical. Therefore, it is inferred that kārmic effects are physical and *mūrta*. Just as the poison of a mouse is *mūrta*, so also *karma* is *mūrta*, because (like the effect of the poison on a mouse) due to experiencing the relation with *mūrta*, such is the fruit of the *mūrta*.

Since the objects are apprehended through the sense organs, which are physical. The kārmic effects, therefore, are also physical and *mūrta*. The law of causality demands that the cause and effect must be mutually reducible. It is thus concluded that since *karmas* themselves are physical and *mūrta*, their effects are also physical and *mūrta*.

134

अरसमरूवमगंधमव्वत्तं चेदणागणमसद्दं।
जाण अलिंगग्गहणं जीवमणिद्दिट्ठसंठाणं॥

arasamarūvamagaṅdhamavvattaṁ cedaṇāgaṇama-
saddaṁ ।
jāṇa aliṅgaggahaṇam jīvamaṇiddiṭṭhasaṅṭhāṇaṁ ॥

Mūrta comes into contact with *mūrta*. Thus, the existing (*mūrta*) kārmic matter enters into a combination or bondage with incoming (*mūrta*) kārmic matter. *Jīva*, though in itself is *amūrta*, is certainly *mūrta* in its worldly existence due to beginningless association with *karma*, absorbs or accommodates those fresh *mūrta karmas* and the *mūrta karmas* accommodate that *jīva* because of the interpenetration.

Commentary by Amṛtacandra

In this world (*loka*), the mundane soul, due to the beginningless association with kārmic matter, comes into contact with fresh *mūrta karmas*. Therefore there is the chance of fresh bondage due to its qualities like attachment, that a kind of bondage with *mūrta karmas.*

Again *jīva*, who is *amūrta* from the internal self-referential point of view (*niścaya naya*) due to attachment, etc. in which beginningless *mūrta karma* is the *nimitta* (supportive factor), accommodates *mūrta karma* in particular (i.e. in which *jīva* and kārmic matter act as *nimitta* of the *pariṇāma* of one another, thereby occupying the same space with the place of *mūrta karma* due to particular relation) and due to the *nimitta* of attachment, etc. *pariṇāma*, the Self obtains knowledge-obscuring, etc. *pariṇāma*, such *mūrta karma* also accommodates *jīva* in a particular form. This is a kind of bondage of the nature of mutually accommodating one another of *jīva* and *mūrta karma*. In this way, *amūrta jīva*'s bondage with *mūrta puṇya–pāpa karma* does not conflict in some respect (*kathñcit*) with one another.

CHAPTER SIX

Āśrava (Stoppage) Padārtha

135

जो खलु संसारत्थो जीवो तत्तो दु होदि परिणामो।
परिणामादो कम्मं कम्मादो होदि गदिसु गदी॥

jo khalu saṁsārattho jīvo tatto du hodi pariṇāmo ।
pariṇāmādo kammaṁ kammādo hodi gadisu gadī ॥

The soul, whose excellent attachment; whose states-of-evolution (*pariṇāma*) that are associated with compassion; and the absence of impure, sinful or debased impulses in his mind, has inflow (*āśrava*) of virtuous, righteous or meritorious (*puṇya*) [*karmas*].

Commentary by Amr̥tacandra

Excellent attachment, transformation of compassion and absence of impurity – these three auspicious psychic dispositions (*śubha bhāvas*) are the causal condition as of the nature of only supportive factor (*nimitta mātra*) of physical (*dravya*) *puṇya āśrava*. Therefore, there are psychic (*bhāva*) *puṇya āśrava* before *dravya puṇya āśrava;* and those *śubha karma pariṇāma* of *pudgalas* are *dravya puṇya āśrava*, which enter through the door of the activities of the mind, speech and body (*yoga*), in which *śubha bhāva* is the supportive factor (*nimitta*).

136

गदिमधिगदस्स देहो देहादो इंदियाणि जायंते।
तेहि दु विसयग्गहणं तत्तो रागो व दोसो वा॥

gadimadhigadassa deho dehādo iṁdiyāṇi jāyaṁte ।
tehi du visayaggahaṇaṁ tatto rāgo va doso vā ॥

Devotion towards the Omniscient Lord Arihanta, the liberated souls (*siddhas*) and the saints (*sādhus*), activity or effort according to righteousness and piety (*dharma*) and faithfully following the preceptors (*gurus*), these are considered excellent attachment (*praśasta rāga*).

Commentary by Amr̥tacandra

This verse is the description of the nature of *praśasta rāga*.

Devotion towards an Arihanta, a *siddha* and a *sādhu*, and activity having primacy of feeling in undertaking *vyavahāra* (external, other-referential) conduct in *dharma*, i.e. righteousness and piety, and following with reverence the preceptors and scholars, that is excellent attachment (*praśasta rāga*), because their subject matter is excellent (*praśasta*).

The *praśasta rāga*, which is primarily devotional, takes place in case of an ignorant person when he has not obtained the upper stage in moral and spiritual development (*guṇasthāna*), also takes place at some time in case of *jñānī*, an intelligent enlightened being, as a condition (*hetu*) to stop attachment, not remaining in a particular place or position and as a condition to eliminate the fever of intense attachment.

137

जायदि जीवस्सेवं भावो संसारचक्कवालम्मि।
इदि जिणवरेहिं भणिदो अणादिणिधणो सणिधणोवा॥

jāyadi jīvassevaṁ bhāvo saṁsāracakkavālammi ।
idi jiṇavarehiṁ bhaṇido aṇādiṇidhaṇo saṇidhaṇovā ॥

If anyone, moved by grief or anguish at the site of the thirsty, the hungry and the miserable offers relief to them, out of pity, then such a behaviour of that person is compassion (*anukampā*).

Commentary by Amṛtacandra

This verse describes the nature of *anukampā*.

After seeing a person suffering from thirst, etc. because of pity and having a desire to offer relief with mental agitation and distress is the *anukampā* of an ignorant person. The *anukampa* of an enlightened (*jñānī*) person, himself happening in the lower stages of moral and spiritual development (*guṇasthāna*), then at the sight of struggling souls in the ocean of worldly existence of births and deaths, there is somewhat regret in mind.

138

मोहो रागो दोसो चित्तपसादो य जस्स भावम्मि ।
विज्जदि तस्स सुहो वा असुहो वा होदिं परिणामो ॥

moho rāgo doso cittapasādo ya jassa bhāvammi ।
vijjadi tassa suho vā asuho vā hodiṁ pariṇāmo ॥

Whenever anger, pride, deceit and greed take refuge in the mind of a *jīva*, they create an agitation or a disturbance (in mind), then the wise call it impure thought.

Commentary by Amṛtacandra

This verse is the description of the impurity of mental thought.

The agitation or disturbance of the mind, due to the coming into operation or rise (*udaya*) of intense anger, pride, deceit and greed, is impurity of thought (*kaluṣata*). The contentment of mind due to the rise of passions like mild anger is freedom from or absence of impurity of the heart (*akaluṣata*). That *akaluṣata* at some time happens in the case of an ignorant soul due to partly destructive and partly subsidential (*kṣayopaśama*) of a particular kind. That which, in the transformation following the rise (*udaya*) of passions like anger, etc. (*kaṣāya*), has displeased the conscious attentiveness (*upayoga*) incompletely, then that (*akaluṣata*) at some time happens in the case of an enlightened person (*jñānī*) in the middle level of *guṇa-sthāna*.

Excellent attachment (*praśasta rāga*), compassion (*anukampā*) and absence of impurity of the heart (*citta-akaluṣata*): these three constitute the inflow (*āśrava*) of virtue or merit (*puṇya*), i.e. *puṇya-āśrava*.

139

सुहपरिणामो पुण्णं असुहो पावत्ति हवदि जीवस्स ।
दोण्हं पोग्गलमत्तो भावो कम्मत्तणं पत्तो ॥

suhapariṇāmo puṇṇaṁ asuho pāvati havadi jīvassa ।
doṇhaṁ poggalamatto bhāvo kammattaṇaṁ patto ॥

Inordinate or excessive carelessness or heedlessness in spiritual matters (*pramad bahula*), impure emotions (*kaluṣata*), hankering for and indulging excessively in sense pleasures, causing anguish to fellow beings and slandering them openly or overtly: these constitute the inflow of sin or vice (*pāpa-āśrava*).

Commentary by Amṛtacandra

This verse is the description of the nature of *pāpa-āśrava*.

The further development of (1) the activity of carelessness (*pramad*); (2) impurity of thought; (3) excessive indulgence in sense pleasures; (4) causing anguish to others; and (5) slandering others: these five inauspicious thoughts (*aśubha bhāvas*) are the causal conditions or the supportive factors (*nimitta mātras*) for physical *(dravya) pāpa-āśrava*. Therefore, there is subjective side of the inflow of sin (*bhava pāpa-āśrava*) before *dravya pāpa-āśrava* and these *aśubha bhāvas* are the *nimitta* of that which are the states-of-evolution (*pariṇāma*) of material *aśubha karmas* entering through the activities of the mind, speech and body (*yoga*), that is *dravya pāpa-āśrava*, i.e. the inflow of objective or physical sin or vice.

140

जह्मा कम्मस्स फलं णिसयं फासेहिं भुंजदे णियदं।
जीवेण सुहं दुक्खं तह्मा कम्माणि मुत्ताणि ॥

jahmā kammassa phalaṁ ṇisayaṁ phāsehiṁ bhuṅjade ṇiyadaṁ ।
jīveṇa suhaṁ dukkhaṁ tahmā kammāṇi muttāṇi ॥

All the four instincts (*saṁjñās*), three *aśubha* thought-points (*leśyās*), subjection to senses, mournful and wicked concentration, undesirable thoughts or knowledge, engaged in *aśubha* activities and infatuation (*moha*): these psychic states (*bhavas*) constitute the spring of sin (*pāpa-prada*).

Commentary by Amṛtacandra

Puṇya and *pāpa* is the result of the four instincts or inanimate feelings of food (*āhāra*), fear (*bhaya*), the sexual instinct (*maithuna*) and the acquisitive instinct (*parigraha*). *Puṇya* and *pāpa* is the result of the three *aśubha* thought-points (*leśyās*) of black (*kṛṣṇa*), blue (*nīla*) and grey (*kapota*), which are the springs of sin or evil (*leśyā* is the activity of the mind, speech and body (*yoga*). They are conciliated or diverted by the rise (*udaya*) of passions like anger; the subjection to senses due to the pre-eminence (*prakarṣa*) of the rise (*udaya*) of attachment and aversion; the joining together of the unpleasant and separation of the pleasant due to being excited by attachment and aversion; the mournful concentration (*ārtta dhyāna*), which is of the nature of release from the pain, anguish and suffering, and desire for a cause of a distressed and depressed state of mind;

wrathful or wicked concentration (*raudra dhyāna*), which has four variations according to the particular inclination of the person towards taking pleasure in violence (*hiṁsanāndi*), lying (*mṛṣanādi*), stealing (*cauryanādi*), and safeguarding of wealth, property, other possessions and objects of sensual pleasures (*viṣayanādi*), due to the *pariṇāma* of cruelty, characterized by anger, passions, knowledge engaged in useless, without any motive, fruitless or unproductive (*niṣprayojana*) work, elsewhere or otherwise than auspicious (*śubha*), *karma* and engaged in *aśubha* action, and lastly imprudent and folly (*aviveka*), is infatuation (*moha*), produced by the rise (*udaya*) of view-deluding (*darśana mohanīya*) and conduct-deluding (*caritra-mohanīya*), i.e. attachment and aversion, *karma*. These give or bestow expansion of subjective conditions of the inflow of sin (*bhāva-pāpa-āśrava*). They also bring about or bestow the expansion of the physical or the objective side of the inflow of sin (*dravya-pāpa-āśrava*), i.e. kārmic material.

CHAPTER SEVEN

Saṁvara Padārtha

141

मुत्तो फासदि मुत्तं मुत्तो मुत्तेण वंधमणुहवदि ।
जीवो मुत्तिविरहिदो गाहदि ते तेहि उग्गहदि ॥

mutto phāsadi muttaṁ mutto mutteṇa vaṅdhamaṇuhavadi ।
jīvo muttivirahido gāhadi te tehi uggahadi ॥

To whatever extent the five senses, anger, etc. for negative emotions, i.e. passions and the four instinctive appetites are restrained or subdued by a person, well-established in the path of righteousness, to such an extent the doorway for the inflow (*āśrava*) of sin is closed for that person.

Commentary by Amr̥tacandra

The path, in reality, is that of *saṁvara*, the stoppage of the inflow (*āśrava*) of sin, which acts as a supportive factor (*nimitta*) in restraining to whatever extent and to whatever period or time, the senses, passions and instincts to such an extent and for that much period, the door of inflow (*āśrava*) of sin is closed.

In the previous verse, it has been said that physical (*dravya*) *pāpa-āśrava* is the condition (*hetu*) of psychic (*bhāva*) *pāpa-āśrava* of the senses, passions and instincts. Here, in this verse, it is to be understood that the restraint of subduing of *bhāva pāpa-saṁvara* is the *hetu* of *dravya-pāpa saṁvara.*

142

रागो जस्स पसत्थो अणुकंपासंसिदो य परिणामो ।
चित्ते णत्थि कलुस्सं पुण्णं जीवस्स आसवदि ॥

rāgo jassa pasattho aṇukhampāsaṁsido ya pariṇāmo ।
citte ṇatthi kalussaṁ puṇṇaṁ jīvassa āsavadi ॥

If a *bhikṣu* (saint), who does not have attachment, aversion or infatuation (*moha*) towards all substances, and who looks upon happiness and misery as the same, then there is no inflow (*āśrava*) of both auspicious and inauspicious (*śubha* and *aśubha*) *karmas*.

Commentary by Amr̥tacandra

One who does not have attachment (*rāga*), aversion (*dveṣa*), or infatuation (*moha*) towards all the other substances, that *bhikṣu* (saint) due to his consciousness being devoid of impurity, looks upon happiness and misery as the same, does not have inflow (*āśrava*) of *śubha* and *aśubha karmas*, but only stoppage (*saṁvara*) of inflow takes place. Therefore, it is to be understood that restraining of *moha*, *rāga* and *dveṣa* is psychic (*bhāva*) *saṁvara*, and that is the supportive factor (*nimitta*) of that which is restraining of material *śubha–aśubha* states-of-evolution (*pariṇāma*) for which the doorway is the activity of mind, speech and body (*yoga*), that is physical (*dravya*) *saṁvara*.

143

अरहंतसिद्धसाहुसु भत्ती धम्मम्मि जा य खलु चेट्ठा ।
अणुगमणं पि गुरूणं पसत्थरागो त्ति वुच्चंति ॥

arahaṁtasiddhasāhusu bhattī dhammammi jā ya khalu ceṭṭhā ।
aṇugamaṇaṁ pi gurūṇaṁ pasattharāgotti vuccaṁti ॥

The saint, who is indifferent or detached, does not, in reality, have virtue or merit (*puṇya*) and sin (*pāpa*) in his activities of mind, speech and body (*yoga*), then *saṁvara* takes place due to the *śubha* and *aśubha bhāva karma.*

Commentary by Amr̥tacandra

In the case of a saint, who is completely detached or restrained from the *puṇya* of the nature of *śubha pariṇāma* and *pāpa* of the nature of *aśubha pariṇāma* in his activities of mind, speech and body (*yoga*), *saṁvara* takes place due to the absence of one's own reason of physical (*dravya*) *karma* due to *śubha–aśubha bhāva*. Therefore, here (in this verse) abstaining from *śubha–aśubha pariṇāmas* (states-of-evolution) is to be understood as the main condition (*hetu*) of *bhāva-puṇya-pāpa saṁvara* and *dravya-puṇya pāpa saṁvara.*

CHAPTER EIGHT

Nirjara (Shedding) Padārtha

144

तिसिदं वुभुक्खिदं वा दुहिदं दट्ठूण जो दु दुहिदमणो।
पडिवज्जदि तं किवया तस्सेसा होदि अणुकंपा॥

tisidaṁ vubhukkhidaṁ vā duhidaṁ daṭṭhūṇa jo du duhidamaṇo ।
paḍivajjadi taṁ kivayā tassesā hodi aṇukaṁpā ॥

That great personality, who after the stoppage (*saṁvara*) of the doorways of *karmas*, both *śubha* and *aśubha*, and is equipped with pure thoughts, practises manifold forms of austerities (*tapa*), will undoubtedly lead to shedding (*nirjara*) of many *karmas*.

Commentary by Amr̥tacandra

Saṁvara, i.e. stoppage of *śubha–aśubha pariṇāma*, and *yoga*, i.e. pure conscious attentiveness (*śuddha-upayoga*), that is equipped with (*saṁvara* and *yoga*), such a person with six external austerities, viz. (1) fasting (*anaśana*); (2) semi-fasting or taking only part of a full meal (*avamaudārya*); (3) voluntarily limiting the range of choice or the variety in the manner of seeking food (*vr̥tti-parisaṅkhyāna*); (4) giving up delicacies or a stimulating diet (*rasa-parityāga*); (5) sitting, sleeping or residing in a lonely place, free from insect afflictions and disturbances (*vivikta-śayyāsana*); and

(6) bodily discomforts (*kāya-kleśa*) and six internal austerities, viz. (1) expiation or atonement (*prāyaścitta*); (2) reverence, veneration or respect (*vinaya*); (3) service (*vaiyyāvṛttya*); (4) study (*svādhyāya*); (5) renunciation or detachment (*vyutsarga*); and (6) concentration or meditation (*dhyāna*). In this way by practising manifold forms of austerities, such a person, in reality, undertakes shedding of many *karmas*. Therefore, here (in this verse, it is said that) the external and internal austerities are capable or powerful enough in the destruction of the strength of *karmas* and due to the growth and development of *śuddhopayoga* (pure conscious attentiveness) is *bhāva nirjara*, and due to its influence, the partial dissociation of the previously acquired *karmas* effects dullness, that is physical (*dravya*) *nirjara*, i.e. shedding of material *karmas*.

145

कोधो व जदा माणो माया लोभो व चित्तमासेज्ज ।
जीवस्स कुणदि खोहं कलुसो त्ति य तं वुद्धा वेंति ॥

kodho va jadā māṇo māyā lobho va cittamāsejja ।
jīvassa kuṇadi khohaṁ kaluso tti ya taṁ vuddhā veṅti ॥

Clinging to the ideal of self-realization, being fortified against *karmas* through *saṁvara*, that saint, who, knowing the Self, regularly meditates upon the pure Self, completely sheds the kārmic dust.

Commentary by Amṛtacandra

That soul, who equipped with the supreme stoppage of the *śubha-aśubha pariṇāma*, knowing the reality of things (what is worth abandoning (*heya*) and what is worth acquiring (*upādeya*)) *tattva* (basic spiritual principles), is eager or ready to practise one's own purpose, whose intellect was stupefied by other purpose, obtaining or achieving the Self, by one's own attainment (or experiencing by one's own self-experience) and due to the identity of qualities (*guṇa*) and the possessor of qualities (*guṇī*) as reality of things, the same knowledge is resolutely mindful or conscious of the Self meditates on the Self, i.e. one's own experience, that *jīva* practising excessively without affection, like pure crystal, undertakes the shedding of kārmic dust amassed previously.

In this verse, it is shown that the main condition (*hetu*) of *nirjara* (shedding of *karmas*) is concentration or meditation (*dhyāna*).

146

चरिया पमादबहुला कालुस्सं लोलदा य विसयेसु ।
परपरितावपवादो पावस्स य आसवं कुणदि ॥

cariyā pamādabahulā kālussaṁ loladā ya visayesu ।
paraparitāvapavādo pāvassa ya āsavaṁ kuṇadi ॥

One, who does not have *moha* (deluded view), attachment and aversion, who has indifference towards the activities of mind, speech and body, it is evident that there manifests the fire of meditation that burns out all *śubha* and *aśubha karmas.*

Commentary by Amṛtacandra

The resolute or unshaken transformation of consciousness in the pure nature of the Self, is real or genuine concentration (*dhyāna*). The manner of appearance of *dhyāna* is being described: when in reality the saint restricts *karmas* (other than one's own non-conscious *karma*), i.e. the maturity of material *karma* of deluded view (*darśana mohanīya*) and attachment and aversion (*caritra mohanīya*). He thereby accordingly serves the conscious attentiveness (*upayoga*), and immerses in an unshaken way that *upayoga* in extremely pure Self alone, that *upayoga*, which is not deluded and is devoid of attachment or aversion, then that saint which is resting or relaxed in his inactive conscious nature and not experiencing mind, speech and body, and does not indulge in his own activity of *karma*. In that case, concentration (*dhyāna*), which is the means of the attainment of supreme endeavour, is revealed. This is like fire that is capable of burning all the rubbish heap of the nature of *śubha–aśubha karmas*.

Amṛtacandra then quotes *Mokṣa Pāhuḍa* 77, which states:

> Even now (in *pañcama kāla*) (in Bhārata Kṣetra) that saint, who is equipped with the purity of triple jewels, i.e. enlightened world-view or right belief, enlightened knowledge and enlightened conduct by meditating on the Self (and concentration on righteousness (*dharma dhyāna*)) gets lordship of heaven or the celestial being (*deva*) of lokāntic. Coming from there one can attain liberation (*mokṣa*).

Amṛtacandra also quotes another verse to the effect:

> There is no end to the revelations and one cannot get across (all) the scriptures, time is short, and we have little or insignificant intellect; therefore, only that is worth learning, by which one can destroy and/or terminate old age and death.

CHAPTER NINE

Bandha (Bondage) Padārtha

147

सण्णाओ य तिलेस्सा इंदियवसदा य अत्तरुद्दाणि ।
णाणं च दुप्पउत्तं मोहो पावप्पदा होंति ॥

saṇṇāo ya tilessā indiyavasadā ya attaruddāni ।
ṇāṇaṁ ca duppauttaṁ moho pāvappadā hoṁti ॥

If the Self, undergoing harmful and agitated states, engages itself in the psychic states (*bhāva*) of *śubha* or *aśubha*, then because of the very same states, the Self gets bound by kārmic matter of various kinds.

Commentary by Amṛtacandra

If, in reality, the Self, by dependence on other, being agitated or harmful (*vikāri*) since beginningless time, due to the coming into operation or rise (*udaya*) of *karmas*, engages in *śubha* or *aśubha bhāva*, then that Self, due to the supportive factor (*nimittabhūta*) of that *bhāva* gets bound by various kinds of material *karmas* (*pudgala karmas*). Therefore, here (in this verse), it is said that *śubha* or *aśubha pariṇāma* (states-of-evolution) due to deluded view (*moha*), attachment and aversion, that is psychic bondage (*bhāva-bandha*). And, these being *nimitta* in the transformation of *śubha* or *aśubha* material *karma*, mutually accommodating each other, that is physical bondage (*dravya-bandha*). *Bandha* is, thus, of two kinds, viz. *bhāva-bandha* and *dravya-bandha*.

148

इंदियकसायसण्णा णिग्गहिदा जेहिं सुट्ठमग्गम्मि ।
जावत्तावत्तेहिं पिहियं पावासवं छिद्दं ॥

indiyakasāyasaṇṇā ṇiggahidā jehiṁ suṭṭhamaggammi ।
jāvattāvattehiṁ pihiyaṁ pāvāsavaṁ chiddaṁ ॥

The *nimitta* of grasping of *karmas* is *yoga*. *Yoga* is the soul's vibratory activities of mind, speech and body. The *nimitta* of kārmic bondage is psychic disposition (*bhāva*). The *bhāva* of the states-of-evolution of the Self (*ātma-pariṇāma*) are equipped with *rati*, i.e. passions and enjoyment of sensual pleasures, attachment (*rāga*), aversion (*dveṣa*) and deluded view (*moha*).

Commentary by Amṛtacandra

Grasping (*grahaṇa*) of the kārmic pattern that flows towards the soul are the manifold kārmic-*skandhas* (aggregates) around *jīva*. This attraction of *skandhas* is due to *yoga* and *jīva* due to the vibrations of the space-points of the Self and the support or dependence of the activities of the mind, speech and body.

Bondage (*bandha*) then is a state of being so bound by kārmic matter around *jīva*, who as *nimitta* is the psychic disposition known as *bhāva*, i.e. *jīva bhāva* of *rati*, *rāga*, *dveṣa* or *moha*-combined *pariṇāma*; in (*vikāra*) produced by the maturity or rise (*udaya*) of deluding (*mohanīya*) *karma*. Therefore, here (in this verse) the external causal condition (*nimitta*) is *yoga* because it is the condition (*hetu*) of the grasping of kārmic matter (*pudgalas*), and the internal causal condition (*nimitta*) is *jīva bhāva* alone because that is the condition (*hetu*) of the particular strength or force (*viśiṣṭa śakti*) and the state, situation or position (*sthiti*) of kārmic matter (*karma pudgalas*).

149

जस्स ण विज्जदि रागो दोसो मोहो व सव्वदव्वेसु ।
णासवदि सुह असुहं समसुहदुक्खस्स भिक्खुस्स ॥

jassa ṇa vijjadi rāgo doso moho va savvadavvesu ।
ṇāsavadi suha asuhaṁ samasuhadukkhassa bhikkhussa ॥

The four thought-constructions (*vikalpas*), i.e. the four external physical (*dravya*) *pratyaya* (influencing factors), [delusion (*mithyātva*), etc. are the causal conditions (*hetu*) of the eight kinds of *vikalpa*, like knowledge-covering, eight kinds of physical *karmas* (*dravya karmas*). But these conditions are themselves due to attachment, and so on *bhāva* (psychic) *pratyayas*. In the absence of these psychic dispositions, there is no bondage.

Commentary by Amṛtacandra

In this verse there is manifestation (or description) of the *mithyātva* and the like, *dravya paryāya*, i.e. modes of material *karma* or *dravya pratyaya*, which are the external conditions of (bondage).[1]

Commenting on this verse Amṛtacandra states:

> In some other Jaina scriptures (*SS* 64-65), the four thought-constructions (*vikalpa*) of delusion (*mithyātva*), lack of moral self-restraints (*asaṁyama*), passions (*kaṣāya*) and the activities of the mind, speech and body (*yoga*) are said to be the causual condition of kārmic bondage (*bandha-hetu*) of knowledge-obscuring, etc. are eight kinds of *karmas*.
>
> The causal conditions (*hetu*) of those *bandha-hetu* are the *jīva bhāva* (the psychic dispositions or mental states of the soul) of attachment, etc. In the absence of attachment, etc. *bhāva*, even when there is a good disposition (*sadbhāva*) of physical (*dravya*)

1 Introductory remark of Amṛtacandra to *Pañcāstikāyasāra* 149.

mithyātva, *dravya asaṁyama*, *dravya kaṣāya* and *dravya yoga*, the souls are not bound.

Therefore, attachment, etc. due to being the internal *bandha hetu*, from the internal, self-referential point of view (*niścaya naya*), this is determined.

Commentary

In *Samayasāra* 109, the four general or universal (*dravya*) *pratyaya*, viz. (1) delusion (*mithyātva*); (2) vowlessness, non-abstinence or lack of moral self-restraints (*avirati*); (3) passions (*kaṣāya* of anger, pride, deceit and greed); and (4) the activities of the mind, speech and body (*yoga*) are said to be the cause of bondage.

Samayasāra 164-165 state:

> *Mithyātva*, *avirati*, *kaṣāya* and *yoga* are the causal conditions of the knowledge-obscuring, etc. eight kinds of *karmas*. The causal conditions of these four (*pratyaya*) are the psychic dispositions of the soul (*jīva-bhāva*) of attachment, aversion, etc., which are *bhāva pratyaya*.

Commenting on *Samayasāra* 164-165, Jayasena observes:

> When on the coming into operation or rise (*udaya*) of *dravya pratyaya* [i.e. *mithyātva*, etc.], if the Self giving up its innate pure nature evolves into attachment, etc. like *bhāva*, then bondage takes place, not by the arising (*udaya*) only. If bondage takes place by *udaya* alone, then there will always be the transmigratory existence in the cycle of births and deaths (*saṁsāra*).

CHAPTER TEN

Mokṣa (Liberation) Padārtha

150-151

जस्स जदा खलु पुण्णं जोगे पावं च णत्थि विरदस्स।
संवरणं तस्स तदा सहासहकदस्स कम्मस्स॥

jassa jadā khalu puṇṇaṁ joge pāvaṁ ca ṇatthi viradassa ।
saṁvaraṇaṁ tassa tadā sahāsahakadassa kammassa ॥

संवरजोगेहिं जुदो तवेहिं जो चिट्ठदे बहुविहेहिं।
कम्माणं णिज्जरणं बहुगाणं कुर्णाद सो णियदं॥

saṁvarajogehiṁ judo tavehiṁ jo ciṭṭhade bahuvihehiṁ ।
kammāṇaṁ nijjaraṇaṁ bahugāṇaṁ kurṇāda so ṇiyadaṁ ॥

In the absence of the causal condition (*hetu*), there is necessarily stoppage (*nirodha*, i.e. *saṁvara*) of the inflow (*āśrava*) of *karmas* in the case of an enlightened person; and in the absence of the *āśrava bhāva* [viz. attachment, aversion and deluded view (*moha*), *SS* 177] there is absence of *karmas*.

Due to the absence of *karmas*, the person becomes *sarvajña*, i.e. all-knowing and all-perceiving Omniscient Lord, and thereby attains the *avyabodha*, i.e. undisturbable or unhindered or uninterrupted state of infinite bliss, which transcends the same feelings, and then there is the absence of feeling of *karma* in the soul and one is placed in the seventh class of Laukāntika heavenly bodies, which is untouched by the sorrows of life.

Commentary by Amṛtacandra

This is the description of the nature of psychic liberation (*bhāva mokṣa*) as of the nature of supreme *saṁvara*, which is, indeed, the causal condition (*hetu*) of physical (*dravya*) *karma mokṣa.*

In reality, the causal conditions (*hetu*) of the inflow of (*āśrava*) of *karmas* are the psychic states (*bhāvas*) of deluded view (*moha*), attachment and aversion of the soul. In the case of the enlightened person, there is absence of them. Due to their absence, there is absence of *āśrava bhāva.* And because of the absence of *āśrava bhāva*, there is absence of *karmas.* Due to the absence of *karmas*, there is only omniscience, all-perceiving and all-knowing, unhindered (*avyabodha*), infinite bliss, transcending the activities of the senses. That is *bhāva mokṣa* of the appellation or designation of liberated state while living (*jīvanmukta*). The explanation of the question "what kind is that?" is given below:

> Here the psychic disposition (*bhāva*), that is meant, intended or implied is initiation in order, succession or sequence of knowing activity of consciousness, covered by the *karmas.* That *bhāva*, in reality, is due to the subjection of the following of coming into operation or rise (*udaya*) of deluding *karma* is impure since beginningless time in the case of the mundane soul and is the condition (*hetu*) of the inflow of physical *karma.* But the very same *bhāva* obtains pre-emminent loss or destruction of the transformation of deluded-view, attachment and aversion in the case of an enlightened person (*jñānī*). Therefore, there is stoppage of *āśrava bhāva* in his case. Due to that, that *jñānī* attains excessive freedom from *vikāra* (damage, altered form, or change for the worse) by the destruction of infatuation (*moha*), then due to being closed the beginningless infinite consciousness and infinite vigour, such a *jñānī* obtains in some respect peak or summit knowledge due to the simultaneous destruction of knowledge-obscuring, the view-obscuring and obstructive *karma*

on the passing by within 48 minutes, the delusionless twelfth stage of moral and spiritual development (*kṣīṇamoha guṇasthāna*) pure knowledge and in this way destruction of his psychic (*bhāva*) *karma* takes place due to the absence of the application or use of the sequence of the nature of his knowing activity. Therefore, due to the absence of *karmas*, he, in reality, always remains glorious, divine (*bhāgyavāna*) Omniscient Lord (*sarvajña*) all-perceiving (*sarva-darśī*) and devoid of the activity of senses, having unhindererd (*avyabodha*) infinite bliss.

Thus, that method of psychic (*bhāva*) *karma-mokṣa* (liberation) and the condition (*hetu*) of physical (*dravya*) *karma-mokṣa* is a manner of supreme *saṁvara*, i.e. stoppage of the inflow (*āśrava*) of *karmas*.

152

जो संवरेण जुत्तो अप्पेट्ठप्रसाधगो हि अप्पाणं।
भुणिऊण झादि णियदं णाणं सो संधुणोदि कम्मरयं॥

jo saṁvareṇa jutto appaṭṭhaprasādhago hi appāṇaṁ ।
bhuṇiūṇa jhādi ṇiyadaṁ ṇāṇaṁ so saṅdhuṇodi kammarayaṁ ॥

The Omniscient glorious divine Lord (Kevali Bhagavān), who is transformed in his nature endowed with complete intuition and knowledge and not having any connection with other substances, such meditation is the condition (*hetu*) of shedding (*nirjara*) of *karmas*.

Commentary by Amr̥tacandra

This verse is the description of the supreme meditation on the Self, which is the causal condition of the shedding (*nirjara*) of *karmas*, which, in turn, is the condition (*hetu*) of physical (*dravya*) *mokṣa*.

In the previous verse, *bhāva-mokṣa* was described. On account of this *bhāva-mokṣa*, the person becomes an Omniscient Lord (Kevalī) endowed with infinite perception and infinite knowledge, with the result of pure knowledge consciousness and due to being suprasensual, that which is devoid of joining together of other substances and due to unshaken resolute pure self-consciousness, such meditation (*dhyāna*), thus, manifests in the purity of self-perception, then it leads to the destruction of previously accumulated *karmas*, i.e. it is the causal condition (*hetu*) of complete shedding (*nirjara*) of *karmas*.

153

जस्स ण विज्जदि रागो दोसो मोहो व जोगपरिकम्मो।
तस्स सुहासुहडहणोज्झणमओ जायए अगणी॥

jassa ṇa vijjadi rāgo doso moho va jogaparikammo ।
tassa suhāsuhaḍahaṇojjhaṇamao jayae agaṇī ॥

When a *jīva*, equipped with *saṁvara* (stoppage of inflow of *karmas*) and obtaining omniscience, succeeding in the shedding of all the remaining *karmas*, then being devoid of pleasant and painful or freed from feeling (*vedanīya*) and age determining (*āyu karma*) and finally physique-making *karma* (*nāma karma*) and family-determining *karma* (*gotra karma*); therefore, that is liberation (*mokṣa*).

Commentary by Amṛtacandra

This verse is the description of the nature of physical (*dravya*) *mokṣa*.

In reality, the Omniscient Lord, after attaining psychic (*bhāva*) *mokṣa* and obtaining the stoppage of the remaining *karma* offspring due to the success of concentration (*dhyāna*), the causal condition of the supreme shedding (*nirjara*) of *karmas*, whose position or situation is somewhat of age-determining (*āyu*) *karma* and keeping in view the measure of *samudghāta* (a sort of overflow from the body for some particular reason), shedding of according to *āyu karma*, at the time of leaving the present condition of mundane existence (*bhāva*) and for attaining the liberated state, the separation of the four *karmas* [known as *aghātiā karmas*], viz. *vedanīya*, *āyu*, *nāma* and *gotra*, that is physical (*dravya*) *mokṣa*.

This ends *mokṣa padārtha*.

Resume (Cūlikā)

154

जं सुहमसुहमुदिण्णं भावं रत्तो करेदि जदि अप्पा।
सो तेण हवदि वंधो पोग्गलकम्मेण विविहेण॥

jaṁ suhamasuhamudiṇṇaṁ bhāvaṁ ratto karedi jadi appā ।
so teṇa havadi vaṅdho poggalakammeṇa viviheṇa ॥

The nature of *jīva* is unimpeded or unrepelled knowledge and perception, which is inalienable with it. In that knowledge and perception, there is an established or appointed existence, which is not blamed, that is spoken of as conduct.

Commentary by Amr̥tacandra

This verse is the description of the nature of the path of liberation.

The established form of right conduct, which is always associated with the innate pure nature of the Self (*jīva*), is the path of liberation. The innate pure nature of *jīva* is, in reality, knowledge and perception because they are inalienable from the nature of the Self. This alienable implies that the nature of *jīva* manifests generically and specifically through knowledge and perception. The existence which is associated with such characteristics and is intrinsic conduct characterized by origination, decay and permanency and due to the absence of attachment, etc., *pariṇāma* (states-of-evolution) is without any blame, that is conduct (*caritra*), the same is the path of liberation (*mokṣa-mārga*).

Conduct in mundane souls is of two kinds: intrinsic, its own self-determined conduct (*sva-caritra*), and extrinsic or other-determined (*para-caritra*), or *sva-samaya* and *para-samaya*. That which is based upon the innate pure nature of the Self is self-determined conduct (*sva-caritra*) and the other-determined conduct is that which is due to the influence of other, i.e. non-Self, is *para-caritra*. Of the two kinds of conduct, that which is existent based upon the innate pure nature of the Self, due to being different from that which is existent in the influence of the other, i.e. the non-Self, is extremely not to be blamed or unblameable, hence it is to be determined as direct *mokṣa-mārga* (path of liberation).

155

जोगणिमित्तं गहणं जोगो मणवयणकायसंभूदो ।
भावणिमित्तो वंधो भावो रदिरागदोसमोहजुदो ॥

jogaṇimittaṁ gahaṇaṁ jogo maṇavayaṇakayasaṁbhūdo ।
bhāvaṇimitto vaṅdho bhāvo radirāgadosamohajudo ॥

Jīva, even when established in his own intrinsic nature (with respect to *dravya*, i.e. substance), if it is associated with accidental or unessential qualities and modes, is called *para-samaya*. If it is established, by being transformed in its own qualities and modes, it is its own (*sva*) *samaya*, then it is able to get itself freed from kārmic bondage.

Commentary by Amr̥tacandra

In this verse, the grasping of *sva-samaya*, which is established in its own intrinsic nature of the Self and by the destruction of *karmas* with the abandoning of *para-samaya* is shown. The mundane soul, even when established unshaken in his own intrinsic nature of knowledge and perception (with respect to *dravya*), when due to being transformed (*pariṇati*) and following the rise (*udaya*) of beginningless deluding (*mohanīya*) *karma* and thereby being stained or tinged (*uparakta*), i.e. impure *upayoga* (conscious attentiveness), then due to grasping of many forms, he has accidental and essential qualities and modes, that is *para-samaya* or *para-caritra*. The same *jīva*, when giving up the *pariṇati* of following the rise (*udaya*) of beginningless deluding (*mohanīya*) *karma*, has extremely pure *upayoga*, then due to grasping of the psychic states of the nature of one (*eka rūpapana*) he becomes established in his own qualities and modes, he is *sva-samaya* or *sva-caritra*.

Now, in reality, by revealing the illumination of the enlightened knowledge and by giving up *para-samaya*, if he grasps *sva-samaya*, then he is necessarily freed from kārmic bondage. Therefore, in reality, to be established in the intrinsic nature of the Self is conduct *mokṣa-mārga* (path of liberation), based on conduct.

156

हेदू चदुव्वियप्पो अट्ठवियप्पस्स कारणं भणिदं।
तेसिं पि य रागादी तेसिमभावेण वज्झंति ॥

hedū caduvviyappo aṭṭhaviyappassa kāraṇaṁ bhaṇidaṁ ।
tesiṁ pi ya rāgādī tesimabhāveṇa vajjhaṁti ॥

Jīva, through attachment, indulges in other instances with auspicious or inauspicious (*śubha* or *aśubha*) psychic disposition (*bhāva*), that the Self, being corrupted or debased from *sva-caritra*, engages in the conduct of *para-caritra*.

Commentary by Amṛtacandra

This verse describes the nature of the initiation in *para-caritra*.

That *jīva*, who under the subjection of the *pariṇati* (transformation) of following, in reality, the rise (*udaya*) of deluding (*mohanīya*) *karma*, and being tinged by attachment, thereby being of impure *upayoga*, acquires *śubha* or *aśubha bhāva* in other substance, that *jīva* is said to be engaged in the conduct of *para-caritra*, being of corrupted *sva-caritra*, because in reality, *pariṇati* in *upayoga* with attachment in other substances is *para-caritra* (other-determined conduct).

157

हेदुमभावे णियमा जायदि णाणिस्स आसवणिरोधो ।
आसवभावेण विणा जायदि कम्मस्स दुणिरोधो ॥

hedumabhāve ṇiyamā jāyadi ṇāṇissa āsavaṇirodho ।
āsavabhāveṇa viṇā jāyadi kammassa duṇirodho ॥

On account of whatever psychic dispositions or states (*bhāvas*) there may flow in merit or virtue (*puṇya*), or sin or vice (*pāpa*) *karma*. Such an *ātman* (Self) on account of such states becomes *para-caritra* (other-determined). So say the Omniscient Lord (Jina).

Commentary by Amṛtacandra

Here (in this verse), conduct that is determined by other things is said to be the condition (*hetu*) of bondage, which blocks the way to liberation.

Here, in reality, the auspicious (*śubha*) mental state is the condition of the inflow (*āśrava*) of merit (*puṇya*) and the inauspicious (*aśubha*) psychic state is the condition of the demerit or sinful (*pāpa*) *āśrava*. On account of both merit and demerit states, the Self is other-determined conduct (*para-caritra*), so the Jina says. Therefore, the activity of *para-caritra* is the path of bondage and not the path of liberation.

158

कम्मस्साभावेण य सव्वण्हू सर्वलोगदरसी य ।
पावदि इंदियरहिदं अव्वावाहं सुहमणंतं ॥

kammassābhāveṇa ya savvaṇhū sarvalogadarasī ya ।
pāvadi iṁdiyarahidaṁ avvāvāhaṁ suhamaṇaṁtaṁ ॥

That *jīva*, which being free from the relations of all others and being inalienable with its own intrinsic nature of knowledge and perception, knows and perceives its own eternal nature to be such, is said to have conduct that is self-determined.

Commentary by Amṛtacandra

Here (this verse) is the description of the nature of that conduct which is self-determined.

That *jīva*, which in reality, due to being devoid of impure conscious attentiveness (*upayoga*), and being free from relations of all others, i.e. abstaining from *upayoga* of other substance, and being inalienable with its own intrinsic nature of knowledge and perception is established, i.e. situated therein, knows and perceives, that *jīva*, in reality, engages in conduct that the self-determined (*sva-caritra*), because the Self is perceived or intuited and understood to be constituted by perception and knowledge, that thought is indeed a self-determined conduct (*sva-caritra*).

159

दंसणणाणसमग्गं ज्झाणं णो अण्णदव्वसंजुत्तं ।
जायदि णिज्जरहेदू सभावसहिदस्स साधुस्स ॥

daṁsaṇaṇāṇasamaggaṁ jjhāṇaṁ ṇo aṇṇadavvasaṅjuttaṁ ।
jāyadi ṇijjarahedū sabhāvasahidassa sādhussa ॥

Whatever, that Self, which is devoid of affective states related to other objects and firmly established in its own intrinsic nature perceives himself to be identical with that free from thought-construction of perception and knowledge, engages therein, that is a self-determined conduct.

Commentary by Amṛtacandra

This verse describes the path of pure *sva-caritra* activity.

That *yogīndra*, which, due to being entirely free from the formation of infatuation (*moha*) and devoid of psychic dispositions of the nature of other substance and being of the intrinsic nature of the Self and following only one, its own intrinsic nature and behaving in an identical way even with its own nature of perception and knowledge with its own intrinsic nature of the Self, that is in reality the conduct of its own behaviour.

In this way, depending on its pure *dravya* and taking recourse to identical ideal and means from the internal, self-referential point of view (*niścaya naya*) the path of liberation has been described, and that which has been described in the earlier verse (verse 107), that which is dependent on the mode of its own and other's condition (*sva-para-hetuka*), taking recourse to different or distinct ideals and means from the external, other-referential point of view (*vyavahāra naya*) had been described. In this it is also not that there is mutual opposition between them because, like golden ore, there is an ideal relationship between *niścaya* and *vyavahāra*. Therefore, one's own glorious, divine (*nija-bhagavān*) application, behaviour or moral conduct (*tīrtha pravartana*) is dependent on both points of view.

160

जो संवरेण जुत्तो णिज्जरमाणोध सव्वकम्माणि ।
ववगदवेदाउस्सो मुयदि भवं तेण सो मोक्खो ॥

jo saṁvareṇa jutto ṇijjaramāṇodha savvakammāṇi ।
vavagadavedāusso muyadi bhavaṁ teṇa so mokkho ॥

Belief in the categories (*padārthas*), such as righteousness and piety (*dharma*) and their respective value, is right belief (*samyakatva*). Understanding the scriptures, which are divided into *aṅgas* and *pūrvas*, is right knowledge. Conduct, according to the twelve kinds of austerities (*tapas*), is right conduct. These three form the path of liberation (*mokṣa-mārga*) from the *vyavahāra* point of view.

Commentary by Amṛtacandra

As a means to *niścaya mokṣa-mārga* (as described in *PKS* 106), this verse describes *vyavahāra mokṣa-mārga*.

The enlightened world-view or right belief, knowledge and conduct is the path of liberation. There, the particular psychic state of belief or faith, which is of the nature of having faith, in the nature of things as they are or "That which is" (*tattvārtha śraddhanam*), that is right belief (*samyakatva*), and given the good disposition of *tattvārtha śraddham*, knowing or understanding of the *padārthas* as described in *aṅgas* and *pūrvas*[1] that is right knowledge and conduct having effort in austerities (*tapa*) of various kinds as described in aphorisms relating to conduct, that is right conduct (*caritra*). Such (*mokṣa-mārga*) is dependent upon *sva–para-hetuka* mode and following different ideals and means according to *vyavahāra naya*, and like the fire set on the golden ore, produces calmness or stability in the pure situations of the innate pure nature of the soul. Although due to the absence of different ideals and means (*bhinna sādhya-sādhya bhāva*), being transformed on its own in pure nature, nevertheless obtains as a means of *niścaya mokṣa-mārga*.

161

जावसहावं जाणं अप्पडिहददंसणं अणण्णमयं ।
चरियं च तेसु णियदं अत्थित्तमणिंदियं भणियं ॥

jāvasahāvaṁ jāṇaṁ appaḍihadadaṁsaṇaṁ aṇaṇṇamayaṁ ।
cariyaṁ ca tesu ṇiyadaṁ atthittamaṇiṅdiyaṁ bhaṇiyaṁ ॥

Whenever the Self (*ātmā*) is in reality contained or collected through the threefold jewels (of right belief, knowledge and conduct) and is identical with them, and neither perceives any other thing nor gives up its own intrinsic nature, then the Self is called as *mokṣa-mārga* from the internal, self-referential point of view (*niścaya naya*).

Commentary by Amṛtacandra

This verse is the description of *niścaya mokṣa-mārga*, the ideal

[1] See J.L. Jaini, 1916, *Outlines of Jainism*, Cambridge: Cambridge University Press.

(*sādhya*) of *vyavahāra mokṣa-mārga*. This verse, together with verse 160, contains the imagination of two kinds of *mokṣa-mārga*: *vyavahāra* and *niścaya*.

This Self in some respect, by the destruction of beginningless ignorance, obtaining *vyavahāra mokṣa-mārga*, and the activity of discriminative knowledge of lack of faith relating to righteousness (*dharma*), etc. and ignorance of *padārtha*-related *āṅga* and *pūrva* knowledge and giving up efforts at austerities, and faith-related *dharma*, etc. and knowledge of *āṅgas* and *pūrvas* and efforts at austerities (*tapa*), and for some reason what is worth grasping is given up and grasping of that which is worth giving up, of the method of retribution (i.e. having the purport of repentance, etc. at whatever time and so long time), due to psychic excellence or effectiveness of the innate pure nature of the Self with right world-view or belief, knowledge and conduct, by being transformed as part and whole and being contained thereby, due to stopping the psychic activity of being void of the thought-construction of giving up and grasping, remains free of being shaken. That time and for that long time, this Self, due to being of established (*niyata*) conduct, is called *mokṣa-mārga* from *niścaya naya*. Therefore, *niścaya mokṣa-mārga* as ideal and *vyavahāra mokṣa-mārga* as ideal and means (*sādhya-sādhana*) are extremely accomplished.

162

जीवो सहावणियदो अणियदगुणपज्जओघ परसमओ।
जदि कुणदि सगं समयं पब्भस्सदि कम्मवंधादो॥

jīvo sahāvaṇiyado aṇiyadaguṇapajjaogha parasamao ।
jadi kuṇadi sagaṁ samayaṁ pabbhassadi kammavaṅdhādo ॥

Whatever the Self, who, in reality, is identical with its own intrinsic nature, perceives, knows and experiences its own Self with the Self and through the Self, that Self alone is identical with the very activity of perception, knowledge and conduct, this is for sure.

Commentary by Amr̥tacandra

This verse is about clarifying the Self's conduct, knowledge and perception.

Whatever the Self, in reality, due to being identical with its own intrinsic nature, experiences that the Self, as its own immediate existence, knows itself, i.e. it illuminates its own and other objects (*sva-para-prakāśaka*), and perceives with it, as it is, that the Self alone is, in reality, *caritra*, *jñāna* and *darśana.* This is definite due to the identity of the causal agent to doer (*kartā*), *karma* and the instrument of that (*kāraṇa*). Therefore, due to conduct, knowledge and perception, the Self whose characteristic is conduct that is established in the intrinsic nature of the Self, such *niścaya mokṣa-mārga* is extremely accomplished.

163

जो परदव्वम्मि सुहं असुहं रागेण कुणदि जदि भावं।
सो सगचरित्तभट्ठो परचरियचरो हवदि जीवो॥

jo paradavvammi suhaṁ asuhaṁ rāgeṇa kuṇadi jadi bhāvaṁ ।
so sagacarittabhaṭṭho paracariyacaro havadi jīvo ॥

Because of the fact that the Self perceives and knows all things without limit, it also experiences infinite happiness on bliss. The *bhavya jīva*, i.e. that which is capable of liberation, knows this fact. The *abhavya jīva*, i.e. that which is not liberable, cannot appreciate that, much less believe or has faith in it.

Commentary by Amr̥tacandra

This verse refutes the capability of all the mundane souls to liberate themselves.

In reality, the reason for bliss is the absence of being contrary or adverseness in the intrinsic nature of the Self. The intrinsic nature of the Self is perception and knowledge. Adverseness in regard to the nature of the Self would be a kind of adverseness to either perception or knowledge. There is the absence of hindrance in perception or knowledge of all in the case of the liberated Self. Therefore, the absence of hindrance naturally manifests in the steady experience of infinite bliss, whose characteristic is being devoid of agitation and distress. In this way, only the *bhavya* knows and accepts this truth; therefore, only that is capable of liberation. The *abhavya jīva* is incapable of having faith in appreciating this truth; therefore, he is incapable or unfit to walk the path of liberation.

Therefore, only some of the mundane souls, not all, are capable of *mokṣa-mārga.*

164

आसवदि जेण पुण्णं पावं वा अप्पणोघ भावेण।
सो तेण परचरित्तो हवदित्ति जिणा परूवंति॥

āsavadi jeṇa puṇṇaṁ pāvaṁ vā appaṇogha bhāveṇa ।
so teṇa paracaritto havaditti jiṇā parūvaṅti ॥

Since *darśana, jñāna* and *caritra* constitute the *mokṣa-mārga,* they are adorned by the wise. If they are determined by other than the Self, i.e. non-Self, they may in a way lead to bondage. However, if they are determined by the Self, then they may lead to *mokṣa* (liberation).

Commentary by Amr̥tacandra

In this verse, by showing that intuition, knowledge and conduct in some respect (say, when associated with attachment, etc.) produces the effect of the condition of bondage. Hence, it is emphasized that if they are associated with the innate pure nature of the Self, then they may directly lead to liberation.

Intuition, knowledge and conduct, even if associated in a small quantity with other things (*para-samaya*), in some respect is also the cause of bondage due to the diffusion in contrary (*viruddha*) action (*kārya*) and its cause (*kāraṇa*). For example, *ghee* (clarified butter), which by nature has a soothing effect, may when associated with fire enhances the heat. And when they (i.e. intuition, knowledge and conduct), abstaining from all the activity of other things (*para-samaya*) [such as attachment] and are associated with the innate pure nature of the Self (*sva-samaya*), then like the release of *ghee* from its association with fire and due to the absence of *viruddha-kārya-kāraṇa bhāva* is the direct cause of liberation alone. Therefore, the activity of *sva-samaya*, which is the conduct in the established intrinsic nature of the Self as the direct cause of *mokṣa-mārga*, is accomplished.

165

जो सव्वसंगमुक्को णण्णमणो अप्पणं सहावेण।
जाणदि पस्सदि णियदंसो सगचरियं चरदि जीवो॥

jo savvasaṅgamukkho ṇaṇṇamaṇo appaṇaṁ sahāveṇa ।
jāṇadi passadi ṇiyadaṁso sagacariyaṁ caradi jīvo ॥

If the Self (*jīva*) through ignorance believes that one attains emancipation from misery due to *śuddha samprayoga*, i.e. auspicious (*śubha*) psychic disposition of reverence or devotion (*bhakti-bhāva*), then the enlightened person (*jñānī*) through ignorance, becomes engrossed in *para-samaya*.

Commentary by Amṛtacandra

This verse provides a description of subtle (*sūkṣma*) *para-samaya*.

The mental activity conciliated with *bhakti-bhāva* towards the Omniscient Lord (Arihanta) and liberated soul (*siddha*), which is the means of fulfilment (*siddhi ke sādhana-bhūta*), is stated here as pure (*śuddha*) *samprayoga*. Due to the frenzy of portion of ignorance, even if an enlightened person (*jñānavāna*) "that *śuddha samprayoga* is the cause of liberation", having regret through such purport, is initiated in (*śuddha samprayoga*), so till then, due to there being good disposition (*sadbhāva*) of a portion of attachment, is called engrossed in *para-samaya*. Then, will not that person be called as being engrossed in *para-samaya*, whose intrinsic mental application is blemished, disgraced or defamed by the blackness of unrestrained or unbridled attachment? Yes, it will necessarily be so called.

166

चरियं चरदि सगं सो जो परदव्वप्पभावरहिदप्पा ।
दंसणणाणवियप्पं अवियप्पं चरदि अप्पादो ॥

cariyaṁ caradi sagaṁ so jo paradavvappabhāvarahidappā ।
daṁsaṇaṇāṇaviyappaṁ aviyappaṁ caradi appādo ॥

The person who has reverence and devotion towards Arihantas, *siddhas*, their images, divine discourses (i.e. the scriptures), the congregation of saints, and knowledge invariably binds a lot of *puṇya karma*. Hence, he can never achieve complete annihilation of *karma*.

Commentary by Amṛtacandra

In this verse, due to being the condition of bondage through *śuddha samprayoga* in some respect, as stated earlier, in that case, *mokṣa-mārga* is prohibited or denied.

The devotion or worship towards Arihanta, etc. due to having *śuddha samprayoga* in some respect and thereby having a portion of attachment and not giving up auspicious (*śubha*) psychic attention (*upayoga*), binds a lot of *puṇya*. But in reality, a person does not annihilate all the *karma*. Therefore, due to being the cause of *para-samaya* application, even a small fragment of attachment is to be given up everywhere or in all cases.

167

धम्मादीसद्दहणं सम्मत्तं णाणमंगपुव्वगदं ।
चिट्ठा तवंहि चरिया ववहारो मोक्खमग्गोत्ति ॥

dharmmādīsaddahaṇaṁ sammattaṁ ṇāṇamaṅga-
puvvagadaṁ ।
ciṭṭhā tavaṁhi cariyā vavahāro mokkhamaggotti ॥

If in the heart of a person there is an iota of attachment towards other or alien objects, then even though he is well versed in all the scriptures, he would still be incapable of knowing and experiencing one's own true pure nature of the Self (*sva-samaya*).

Commentary by Amṛtacandra

In this verse, it is shown that in the absence of obtaining *sva-samaya*, attachment is the condition (*hetu*).

If in the heart of a person, there is present even a small fragment of a particle or grain of attachment, that person may be well-versed in all the moral and philosophical doctrines and principles, still he does not, in reality, experience the pure nature of *sva-samaya* free from the darkening effect of attachment. Therefore, by preserving the rule of removing the cotton sticking to the bow used for scritching or carding cotton (*dhunakī*), it is worth removing for the Self gradually, or by due stages, the particle or grain of attachment concerning even the Arihanta, etc. as the condition (*hetu*) of the well-known *sva-samaya*.

168

णिच्चयणयेण भणिदो तिहि तेहिं समाहिदो हुजो अप्पा ।
ण कुणदि किंचिवि अण्णं ण मुयदि सो मोक्खमग्गोत्ति ॥

ṇiccayaṇayeṇa bhaṇido tihi tehiṁ samāhido hujo appā ।
ṇa kuṇadi kiṅcivi aṇṇaṁ ṇa muyadi so mokkhamaggotti ॥

To whomsoever there is no ability to keep in check the wanderings of the mind due to the good disposition of attachment, there is no possibility for the Self in preventing or stopping the auspicious and inauspicious (*śubha* and *aśubha*) *karmas.*

Commentary by Amṛtacandra

Here (in this verse), there can be in reality no devotion to Arihanta, etc. without attachment. Due to further development of attachment, it is not possible for the Self to continue to be devoid of the expansion of the thought-constructions and because of that, there is no stoppage of *śubha* and *aśubha karmas*. Therefore, this is only the sensuousness or flirtations of the anguish or distress of the nature of attachment, which is the cause of continuous succession of senseless acts.

169

जो चरदि णादि पिच्छदि अप्पाणं अप्पणा अणण्णमयं।
सो चारित्तं णाणं दंसणमिदि णिच्चिदो होदि ॥

jo caradi ṇādi picchadi appāṇaṁ appaṇā aṇaṇṇamayaṁ ।
so cārittaṁ ṇāṇaṁ daṁsaṇamidi ṇiccido hodi ॥

Therefore, the Self, who is desirous of liberation (*mokṣa*) must be without attachment (*niḥ saṅga*) and sense of mineness (*nirmam*), and then is devoted to the liberated souls (*siddhas*) [i.e. his own true nature which is of the same (intrinsic) nature as *siddha*, *NS* 49], thereby he attains liberation or salvation (*nirvāṇa*).

Commentary by Amṛtacandra

This verse is the exposition of the complete destruction of the anguish of the nature of attachment, which is worth doing.

Due to the development of attachment, etc. there is wandering of the mind and because of that there is kārmic bondage. This is stated earlier. Therefore, one, who is desirous of *mokṣa*, the cause of *karma-bandha*, i.e. the wanderings of the mind, having its root in the development of attachment is worth absolutely and completely destroying. By destroying that without any remainder, the *niḥ-saṅgatā* and *nirmamatā* become well known, as the Self preserving the calmness of the pure intrinsic nature of the Self and the supreme state of devotion to *siddha* is well known as regards the activity of *sva-samaya*. The Self therefore attains spiritual fulfilment or accomplishment by completely destroying without any remainder (*niḥśeṣa*) bondage of *karma*.

170

जेण विजाणदि सव्वं पेच्छदि सो तेण सोक्खमणुहवदि।
इदि तं जाणदि भविओ अभव्वसत्तो ण सद्दहदि॥

jeṇa vijāṇadi savvaṁ pecchadi so teṇa sokkhamaṇuhavadi ।
idi taṁ jāṇadi bhavio abhavvasatto ṇa saddahadi ॥

One's intellect may be inclined towards the *tīrthaṅkaras* and the nine principles (*tattvas*) or *padārthas* (categories). One may have an interest in the affection to the scriptures. One may have self-control and austerities. With all these, one is not capable of realizing his own true Self. In the case of such a person, emancipation (*nirvāṇa*) is remotely situated.

Commentary by Amr̥tacandra

In this verse, it is shown that the activity of *para-samaya* of the nature of devotion to Arihanta and the like, due to the absence of direct condition of liberation (*mokṣa*), is said to have good disposition of being the condition of liberation traditionally.

Even though a person in reality is active, energetic and engaged in the condition of liberation and attaining unimaginable or inconceivable self-control and austerities, due to not having such divine strength as capable of ascending or rising to the rule of supreme detachment, like the rule of cotton sticking to the bow used for carding or scritching cotton (*dhunakī*), cannot give up the activity of *para-samaya* of having interest (affection) in nine *tattvas* or *padārthas* and Arihanta, etc. that *jīva* in reality does not obtain direct liberation. He attains liberation traditionally by securing the happiness (in fact, the anguish and distress) of heaven.

171

दंसणणाणचरित्ताणि मोक्खमग्गोऽति सेविदव्वाणि ।
साधूहि इदं भणिदं तेहिं दु वंधो व मोक्खो वा ॥

daṁsaṇaṇāṇacarittāṇi mokkhamaggo 'ti sevidavvāṇi ।
sādhūhi idaṁ bhaṇidaṁ tehiṁ du vaṅdho va mokkho vā ॥

The person, who has reverence or devotion to *Arihants*, *siddhas* (liberated souls), their images (*caitya*) and the divine discourses (*pravacana*) or the scriptures, engages in supreme self-control and austerities [yet so long as he is incapable of apprehending the true greatness of the Self on account of attachment to external ideals], will secure the world of celestial beings.

Commentary by Amr̥tacandra

This verse is the exposition of the obstruction of direct *mokṣa-mārga* due to only attachment and devotion to Arihanta, etc.

The person, who in reality engages oneself with intellect under the subjection of devotion to Arihant and so on, burns in austerities consisting of predominantly supreme self-control, that *jīva* with that much anguish or distress of the nature of attachment, thereby his inner soul is tarnished. His internal heart is infatuated with the pleasures of the poisonous tree of sense objects. Such a world of heaven, which is obstructive of direct *mokṣa*, acquiring that for a very long period, being burnt with burning coal of the nature of attachment, is distressed or tormented with suffering being burnt internally.

172

अण्णाणादो णाणी जदि मण्णदि सुद्धसंपओगादो ।
हवदित्ति दुक्खमोक्खं परसमयरदो हवदि जीवो ॥

aṇṇāṇādo ṇāṇī jadi maṇṇadi suddhasaṁpaogādo ।
havaditti dukkhamokkhaṁ parasamayarado havadi jīvo ॥

Therefore, the Self, desirous of *mokṣa* should not have even a little attachment. By doing that, the liberable (*bhavya*) *jīva*, by being non-attached (*vītarāga*). He is able to cross the ocean of *saṁsāra*, the transmigratory existences of births and deaths.

Commentary by Amṛtacandra

This verse is the conclusion of the essence of the direct *mokṣa-mārga* of the nature of the purport of the scriptures.

In reality, the leading ingredient of the direct *mokṣa-mārga* is the state of being non-attached (*vītarāgatā*). Therefore, in reality, even the attachment to Arihanta, etc. [alien persons deserving worship] is something like burning sandalwood. [It may smell sweet, nevertheless it will burn.] The subtle attachment (to external ideals) may lead to happiness (which indeed is anguish and distress), i.e. affliction (*kleśa*) of the world of heaven will scorch the soul. Hence, the great person, desirous of direct *mokṣa*, must rid himself of all kinds of attachment, must become completely non-attached (*vītarāgī*). Then one will be able to soon attain emancipation (*nirvāṇa*) by immersing in the ocean of supreme nectar of the innate pure nature of the Self by crossing over to the ocean of transmigratory existences of births and deaths (*bhava-sāgara*), which is immensely terrible as in that the waves of pleasure and pain are boiling and splashing up, and which is excited, agitated and bubbling up burning with the fire of *karmas*.

Victory to the *vītarāgatā* (complete non-attachment), which is the purport (*tātparya*) of the scriptures by being the essence of direct *mokṣa-mārga*.

Tātparya (purport) is of two kinds: (1) *sūtra-tātparya*, and (2) *śāstra-tātparya*.

The purport of this aphorism (*sūtra*) (is not only the purport of this verse), it is also the purport of the whole book, *Pañcāstikāyasāra*. (The exposition of *tātparya* of the *sūtra* is given in each verse.)

The exposition of *mokṣa* principle (*tattva*), which is the essence of all the efforts or worth-achieving purposes (*puruṣārthas*) [which are for righteousness (*dharma*), satisfaction of economic needs (*artha*), taking care of one's physical and instinctive needs of food, sex (*kāma*), etc. and realization of the spiritual happiness of liberation (*mokṣa*). Of these, *mokṣa* is the highest objective of human endeavour].

The five *astikāyas*, the inherent condition of expounding the principle (*tattva*) of *mokṣa*, the essence of all *puruṣārthas*, and through the exposition of the six substances (*dravyas*), the nature of all the things has been shown, and by the detailed description of the nine *tattvas* or *padārthas*, in which are revealed thought-construction of (difference) of bondage (*bandha*) and liberation, the locale of *bandha-mokṣa* relating to *bandha-mokṣa*, and which has proper investigation or examination of *niścaya* and *vyavahāra mokṣa-mārga* and the cause of direct *mokṣa*, of which supreme non-attachment is the entire heart – such as this real, genuine or true glorious, divine scripture, i.e. *Pañcāstikāyasāra*, of which non-attachment alone is the purport from the supreme state of reality (*paramārtha*).

Therefore, one should certainly understand the compatibility in the harmony between the practice of *vyavahāra* and *niścaya* paths, and thus realize the ideal of non-attachment (*vītarāgatva*) as the immediate and direct cause of *mokṣa* (liberation). Only by the practice of *vyavahāra* and *niścaya* paths being complimentary, not contradictory or opposed to each other (*avirodhena*), one can attain the objective of non-attachment (*vītarāgatā*), not otherwise (*na punar anyathā*).

The above reality is now being understood in detail.

Due to beginningless intellect of viewing or understanding things by differentiating them, the primary or elementary *jīva* descends on the *tīrtha*, i.e. the path of liberation by the dependence or support of differentiating the ideal and the means through the external, other-referential point of view (*vyavahāra naya*).

For example, (1) it is worth your faith; (2) this is not worth your faith; (3) this is a believing, devout or reverential person; (4) this is faith or reverential belief (*śraddhān*); (i) this is knowable; (ii) not knowable; (iii) this is the knower; and (iv) this is knowledge; (a) this is to be performed; (b) this is not to be performed; (c) this is the performer of conduct; and (d) this is conduct or behaviour. In this way, (1) this is to be done, duty or task; (2) this is not to be done; (3) this is the causal agent or doer (*kartā*); and (4) this is the action, an act or deed (*karma*).

Thus, by looking or viewing divisions, sections or parts, in which there is joyful enthusiasm; in this way the primary or elementary persons tear up or detach infatuation, i.e. attachment, etc. At some point of time, due to ignorance, pride and negligence, the power of the Self becoming loose or slack, they make use of atonement and again and again, they always remain energetic and industrious in differentiating ideal and means (*bhinna-sādhya-sādhana*), thereby attaining, some purification. Due to the absence of *bhinna-sādhya-sādhana bhāva*, they depend on *niścaya naya* (internal, self-referential point of view), whose nature is non-differentiating (*abhedapana*), which contains or in which are collected intuition, knowledge and conduct together and dire to the absence of all ostentatious or shows of dire actions, that which is unwaveringly of supreme consciousness and which is prosperous with extreme or abounding happiness, having calmness in such a glorious, divine Self, there arises successively equable or equitable, harmonious *bhāva* (psychic disposition); therefore, after attaining supreme non-attached (*vītarāga*) *bhāva*, one experiences direct liberation (*mokṣa*).

Now, those who depend upon or take shelter only on *vyavahāra* (*kevala-vyavahāra-lambī*), such an ignorant persons' initiation and its result is being described below:

Those who are *kevala-vyavahāra-lambī*, they, in reality, by looking at *bhinna-sādhya-sādhana bhāva*, always obtain excessive regret in three ways:

1. again and again, due to their mind being attached to the *adhyavasāna*, i.e. the impure psychic states of modes, such as delusion (*mithyātva*), that are superimposed on the Self due to coming into operation or rise (*udaya*) of material *karmas*, of faith in righteousness, piety (*dharma*), and so on
2. due to many kinds of net thought-constructions (*vikalpa*) and their inborn concepts or ideas as a result of learning scriptures, the conscious activity is therefore variegated, unusual or peculiar, and
3. due to the initiation of the collection of various kinds of conduct and austerities and great merriment or uproar of *kriyā-kāṇḍa*, (unfortunate or dire activities or incidents).

They have sometimes interest in some sense objects and sometimes have *vikalpa* in other objects, sometimes behave in some conduct, for the sake of *darśana*, at some time they are quiet [i.e. *praśama* subsidence of passions]. At some other time, they are enthusiastic (*saṁvega*) about something. At some time, they observe compassion (*anukampā*). At some time, they maintain or preserve *astikāya*, i.e. firm conviction in the divinity of every living being. These are the prerequisites of enlightened world-view or right belief, *samyak darśana*]. They remain always prepared as a condition (*hetu*) of stopping the rise of doubt or scepticism (*śaṅkā*), desire or expectations (*ākāṅkṣā*), contempt or disgust (*vicikitsā*), the delusive notions or follies and contemplating again and again for increasing with enthusiasm the cultivation of virtuous dispositions in oneself, re-establishing those faltering on the path of righteousness (*sthiti karaṇa*), disinterested affection (*vātsalya*) and propagation of the values of life (*prabhāvanā*) [the eight characteristics of enlightened world-view or right belief (*samyak darśana*)]. They study for practising knowledge. They engage themselves in difficult tasks to hold religious observances or austerities (*upādāna*) for the expansion of *vinaya*, i.e. reverence, internal penance, ethical code of conduct and modesty. They ward off the fault of concealment or heresay and are extremely careful in the purification of the object of knowledge (*artha*), sense-object (*vyañjanā*) and both (*tadubhaya*). For practising conduct, they are engaged in the five great vows of being completely indifferent to violence, untruth, stealing, unchastity and acquisitiveness. They are active in restraining the activities of mind, speech and body properly (*samyak yoga-nigraha*), is the characteristic of which, such *guptī* (restraint of mind, speech and body); are united or combined with five *samiti*, i.e. carefulness in walking, speaking, eating, lifting and keeping things, and cleanliness.

They are always enthusiastic in the six external austerities, viz. (1) *anaśana* (fasting); (2) *avamaudārya* (semi-fasting, reduced diet or taking only a part of the full meal); (3) *vr̥ttiparisaṁkhyāna* (voluntarily limiting the range of choice or the variety and the manner of seeking food); (4) *rasatāparityāga* (giving up delicacies or stimulating diet); (5) *vivikta-śayyāsana* (sitting, sleeping or residing in a lonely place, free from insect afflictions and all disturbances); and (6) *kāya-kleśa* (bodily discomforts).

They keep restraint or control over the mind through the six internal austerities, viz. (1) *prāyaścitta* (expiation or atonement); (2) *vinaya* (reverence, veneration or respect); (3) *vaiyyāvr̥tya* (service); (4) *vyatasarga* (renunciation or detachment); (5) *svādhyāya* (study); and (6) *dhyāna* (concentration or meditation). They remain astray with all force from *karma-kāṇḍa* (useless practices) for increasing the vigour, by doing that due to the primacy of consciousness of one's own activities (*karma-cetanā*), although they have excessively prevented the inauspicious (*aśubha*) *karma* tendency and have grasped quite well the auspicious (*śubha*) *karma* tendency, crossing over the ostentation or pretension of *kriyā-kāṇḍa*.

They wander in the transmigratory existences for excessively long time by obtaining traditionally the anguish or distress (*kleśa*) of the celestial world, etc. by having slow mental process due to the weight of virtue or merit (*puṇya*) by not producing or giving rise to even a little knowledge consciousness (*jñāna cetanā*), the state of evolution of unity or oneness of *darśana, jñāna* and *caritra* (intuition, knowledge and conduct). It is also said: "Those who have primacy in conduct, they being devoid of supreme reality of the nature of *sva-samaya*, do not experience the essence of *niścaya* pure self of conduct."

Now, the initiation and result of those who depend upon or take shelter only on *niścaya* (*kevala niścaya-avalambī*) (such *ajñānī*, i.e. ignorant persons) is being described below:

Now, those who are *kevala-niścaya-avalambī* (*ajñānī*) persons, by behaving with intellect are alienated from all the ostentation of *kriyā–karma-kāṇḍa*, remain by imagining whatever produces pleasure.

a. They bind themselves with sin (*pāpa*) by the following activities: censuring and reproaching or disdaining *bhinna-sādhya-sādhana-bhāva* and not obtaining or achieving non-differentiating (*abhinna*) ideal or objective and means (*sādhya-sādhana*) *bhāva*, in the interval (in the inauspicious (*aśubha*) condition alone, apart from auspicious (*śubha*) and pure (*śuddha*);
b. being lazy, filled with the intoxicating drink (wine or alcohol),
c. frenzied or lustful, like sleeping as fainting, or swooning, i.e. insensibility;
d. satisfied by eating a lot of *ghee*, sugar, i.e. a dish of rice boiled in milk with sugar (*khīra*), due to such fat, bulky or

heavy body, and has arisen *jaḍatā*, i.e. lifelessness, inertness, or insensibility (inactivity) and stupidity due to fearsome, terrible or painful *buddhi-bhraṣṭa*, i.e. corrupted or debased intellect, as if the particular consciousness is closed,

e. not adopting consciousness of one's activity (*karma-cetanā*) due to the fear of the bondage of merit (*puṇya-bandha*), and not obtaining supreme calmness of knowledge consciousness (*jñāna-cetanā*); and

f. behaving under the subjection of only negligence (*pramāda*), manifested or not-manifested (expressed or non-expressed) having the tendency of the primacy of consciousness of experiencing merely the fruits of one's past *karma* (*karma-phala-cetanā*). Amṛtacandra here quotes an old verse:

> Depending on *niścaya naya*, but not knowing [sense, purport or] the reality of *niścaya naya*, some living beings, by being lazy and external conduct, destroys the condition of conduct.

Now, an enlightened person's (*jñānī*) initiation according to the role that there is combination, concord or a harmony between *niścaya naya* and *vyavahāra naya* and its result, is described below:

> But those glorious divine great persons, who constantly make efforts for liberation, not clinging one-sidedly to only *vyavahāra-avalambī* or only (*kevala*) *niścaya-avalambī*, but being extremely *madhyastha*, i.e. indifferent to either of them [i.e. understanding the two paths properly, along with the compatibility and harmony between the two paths] and aspiring towards the relaxation in the Self principle (*ātma-tattva*) of the innate pure (nature of) consciousness, to avoid the tendency of giving rise to negligence (*pramāda*) and being extremely indifferent to the activity of *kriyā-kāṇḍa*, they always remain worthy of experiencing the Self through the Self in the Self. Such glorious divine (*bhagavanta*) great persons, in reality, according to the relaxation in one's own spiritual principle (*sva-tattva*), renouncing successively the

actions (*karma-saṁnyāsa*), being extremely devoid of *pramāda* and being excessively unshaken (*niṣkampa*) like a plant, they have excessively destroyed the experience of *karma-phala* (fruits of *karmas*), thereby being of dampened of ardour or energy towards kārmic experience, arisen only by knowledge experience, extremely filled with tāttvic *ānanda*, the bliss of spiritual principles, they are soon the enjoyer of the happiness of liberation of the fruit of eternal revealed word.

Commentary by Jayasena

Jayasena states that liberation is attainable only when both *niścaya* and *vyavahāra* points of view are *paraspara sāpekṣa*, i.e. mutually related and complementary to one another, not when they are *nirapekṣa* (unrelated). He refutes the views of depending one-sidedly on either *vyavahāra* or *niścaya* point of view.

173

अरहन्तसिद्धचेदियपवयणगणणाणभत्तिसंपण्णो ।
वंधदि पुण्णं बहुसो ण दु सो कम्मक्खयं कुणदि ॥

arahantasiddhacediyapavayaṇagaṇaṇāṇabhattisaṁpaṇṇo ।
vaṅdhadi puṇṇaṁ bahuso ṇa du so kammakkhayaṁ kuṇadi ॥

As a causal condition (*hetu*) of propagation of *mokṣa-mārga*, this work by the name of *Pañcāstikāyasāra-saṁgraha* aphorisms (*sūtra*), containing the essentials of divine discourses, is composed by me as motivated, stimulated or encouraged by the devotion towards the divine revelation.

Commentary by Amṛtacandra

This verse is indicative of the conclusion of the completion of the promise of Kundakunda.

That which is skilful in producing supreme freedom from worldly desires (*vairāgya*), which has the name of the commandment of the precept of the glorious divine Omniscient Lord, the propagation of which or production by pre-eminent transformation, the condition of the propagation of the path (*mārga*) [of liberation], due to unsteady mind owing to the flow or speed of affection of the supreme scripture (of divine revelation), which is in the form of aphorisms of the instructions of the Omniscient Lord, of which the short summary of the exposition by being the causal agent or doer (*kartā*) of the real, genuine or true sense of the nature of all the things being very much extensive or detailed and yet is the essence of the divine discourses.

174

जस्स हिदयेणुमत्तं वा परदव्वं हि विज्जदे रागो।
सो ण वजाणादि समयं समस्स सव्वागमधरो वि ॥

If in the heart of a yogī there is an iota of affection towards alien objects (even in an arahanta), then though he be well versed in all the scriptures, he would be incapable of appreciating his true Self (Paramātmā) through immediate experience.

Commentary by A. Chakravarti

Devotion to external ideals may be useful up to a certain stage of development. The ideal of *vītarāga* is distinctly related to the Self. Absolute freedom of the Self from attachment is the meaning of *vitarāga*. Complete absence of *rāga* is certainly inconsistent with attachment and devotion to an *arahanta* or a *siddha*. In so far as the Self is concerned, such ideal personalities are external. The right ideal must be to attain oneself the state of perfect *siddha*. This realization of *siddha* state cannot be reached by having attachment to external objects. Kundakunda, therefore, emphasizes the importance of efforts to transcend even this subtle kind of attraction however worthy the object of the ideal may be. The Self and the Self alone must be the final and the ultimate ideal of spiritual evolution. To have *bhakti*, which is a subtle attraction and then to believe that somehow the final good will be *mokṣa*, is to forget the importance of self-effort towards self-realization. According to Kundakunda, the traditional religious reverence is only a stepping stone for the higher spiritual goal. He who shrinks from transcending the traditional orthodoxy is unfit to live on a higher plane. He is infected with the delusion that what is only a means is an end in itself. That, what is merely relative, is the absolute. That, what is of temporary value, has eternal value.

175

घरिदुं जस्स ण सक्कं चित्तुब्भामं विणादु अप्पाणं ।
रोधो तस्स ण बिज्झदि सुहासुहकदस्स कम्मस्स ॥

To whomsoever who cannot bear the turbulence of the mind apart from and independent of the pure contemplation of the Self, there is no possibility for such a person to prevent *karmas* born as a result of pleasure and pain.

Commentary by A. Chakravarti

There can be no devotion to an *Arahanta* and others apart from an affective state. When there is affection, an agitated mind cannot be avoided. When the heart is agitated, the Self loses its calmness and if the calmness of the Self is disturbed, it is not possible to prevent the appearance of pleasure and pain. As a result, there is no escape from happiness and misery. There can be no emancipation from *saṁsāra*. Affection, even if it is meagre, is therefore the cause of this series of undesirable events.

176

तह्मा णिव्वुदिकामो णिस्संगो णिम्ममो य हविय पुण्णो ।
सिद्धेसु कुणदि भत्तिं णिव्वाणं तेण पप्पोदि ॥

One who seeks to attain *mokṣa* must completely eradicate every kind of desire from the heart.

Therefore, a person who aims at *nirvāṇa* must not say "I love this, this is mine". If one does this, one can very well be devoted to his own true Self which is the liberated state of a *siddha*. A person will attain *nirvāṇa* through that same self-realization.

Commentary by A. Chakravarti

The great soul that understands the truth of the causal sequence stated earlier from desire to an agitated mind, and from thence to *karma,* and finally to *saṁsāra,* struggles to escape from the attraction of the environment and from the idea of ownership of alien things. When the relation of the Self to alien things either through attraction or through owning is severed, then there is no possibility of desire springing up in the heart. No desire means no spiritual discord. Spiritual harmony repels *karmas*. Prevention of *karmas* means disappearance of *saṁsāra*. The absence of *saṁsāra* certainly implies its opposite, i.e. *mokṣa*. Hence a person who longs for spiritual perfection and kārmic emancipation must never say "This I love, This is mine".

177

सपयत्थं तित्थयरं अभिगदबुद्धिस्स सुत्तरोइस्स।
दूरतरं णिव्वाणं संजमतवसंपओत्तस्स॥

One may understand the true nature of a *tīrthaṅkara,* who is the basis of the nine *padārthas*. One may have interest in and devotion to the scriptures. One may have self-control and practise penance. However, if one is not capable of realizing his own true Self, *nirvāṇa* is beyond reach for such a person.

Commentary by A. Chakravarti

Even though a person is well versed in Āgamas and well acquainted with the nature of reality, if he is replete with devotion and worship to objective ideas, he will only secure the happiness of *svarga* as the immediate fruit thereof. *Nirvāṇa* or *mokṣa* is very remote and a person can obtain that only by a tedious process of spiritual development. The immediate and direct antecedent of *mokṣa* is the realization of the true Self. Worship of and devotion to any other ideal, however worthy it may be, is only a circuitous path to reach the goal.

178

अरहंतसिद्धचेदियपवयणभत्तो परेण णियमेण।
जो कुणदि तवो कम्मं सो सुरलोगं समादियदि॥

The person who has love and reverence towards *arahanta*s and *siddhas*, their representations and the Divine Word or Āgama, however well he is engaged in the best form of *tapas*, so long as he is incapable of apprehending the true greatness of the Self on account of attachment to external ideals, will inevitably through his efforts enter only the world of the *deva*s.

Commentary by A. Chakravarti

Whether now or hereafter, the direct and immediate condition of *mokṣa* is the apprehension of the true Self and not attachment to external ideals. The person who has not grasped the Self through all his efforts, associated with worship and reverence, will only secure the happiness of the *deva*s. This is merely a higher state of *saṁsāra*, not *mokṣa.*

179

तह्मा णिव्वुदिकामो रागं सवत्थ कुणदि मा किंचि।
सो तेण वीदरागो भविओ भवसायरं तरदि॥

Because of this truth that even the subtle attachment to external ideals hinders *nirvāṇa*, let not the person who desires *mokṣa* have any kind of attachment to things, living or non-living, desirable or undesirable, then he becomes the *vitarāga,* the non-attached. He will thereby be able to sail across the ocean of *saṁsāra.*

Commentary by A. Chakravarti

The essential ingredient of the direct and immediate path to *mokṣa* is the state of *vitarāga* (non-attachment). Attachment to alien persons and things deserving worship is something like burning sandal wood. It may smell sweet, but it will nevertheless burn. The subtle attachment to external ideals may lead to happiness but that very happiness will scorch the soul. Hence a person, who yearns for freedom from *saṁsāra* must get rid himself of any kind of attachment, he must become a *vitarāgī.* He will then be able to conquer the misery of life and inherit the kingdom of *nirvāṇa.*

This is the meaning of this verse and the same is the purport of *Pañcāstikāyasāra*, which seeks to understand the nature of Jineśvara (the great victor), the all-knowing of infinite qualities and the *vītarāga* (non-attached).

The *Pañcāstikāyasāra* is called *prabhūta* (well filled with the meaning) because it deals with the nine *padārthas* and thus describes the nature of the systems of reality. Since it deals with *dravyas* or the existences, it is also called *dravya-niyoga.* It clearly explains the five *astikāyas*, the six *dravyas*, the seven *tattvas* and the nine *padārthas.* It also examines the nature of bondage and of the bound, and of liberation and the liberated.

Lastly, in the Cūlikā (Appendix), there is a clear distinction between the absolute path and the relative path as well as indications that non-attachment is the direct and immediate condition attaining *nirvāṇa.* Non-attachment and perfect peace of the heart is the central doctrine emphasized by an embodied in this *Pañcāstikāyasāra.* This *vītarāgatva* (state of non-attachment) is only attained by the right believer who appreciates the comparative merits of the two paths, relative and absolute *vyavahāra* and *niścaya,* but not to one who is incapable of appreciating the various aspects of truth and who clings to only a particular and fractional aspect as the whole truth, i.e. one who believes in the *ekānta* aspect of reality and is incapable of appreciating the ideal of *vītarāga*, and the difference between the absolute path and the relative path. He is far away from the goal. The true believer will certainly understand the compatibility and the harmony between the two paths and thus will seek to attain the ideal of non-attachment (*vītarāgatā*) as the immediate condition of *mokṣa*.

180

भग्गप्पभावणट्ठं पवयणभत्तिप्पचोदिदेण मया ।
भणियं पवयणसारं पंचत्थियसंगहं सुत्तं ॥

The *Pañcāstikāyasāra*, which contains the essentials of divine revelation, has been composed by Kundakunda in order to explain the *mokṣa-mārga*. It is motivated by devotion (*bhakti*) towards the Divine Word.

Commentary by A. Chakravarti

Kundakunda explains the purpose of writing the *Pañcāstikāyasāra*. The *mokṣa-mārga*, or the path, is revealed by a Jina. But divine revelation (*śāstra*) is elaborate and extensive. In order to show the path to the believers, this short summary of the Divine Word is written with the name of *Pañcāstikāyasāra*.

[Thus ends Book II of the *Pañcāstikāyasāra*.]

Appendix

Distinguishing Features of the Pañcāstikāyasāra and Other Works of Kundakunda

Pañcāstikāyasāra

The term *pañcāstikāyasāra* consists of two words, viz. *pañcāstikāya* and *sāra*. As the name suggests, the *Pañca-asti-kāya-sāra* (*PKS*) deals with the essence (*sāra*) or the essential nature of *pañca* (five) *astikāya* meaning entities, such as *jīvas* (souls or living beings), etc. mentioned (in *Pañcāstikāyasāra* 4 that are capable of occupying space (called *kāya*)).

The term *kāya* (corporeals) implies relation to many spatial points. Existences that can be so related to space are called in Jainism, *astikāya*, which too consists of two words, viz. *asti*, i.e. that has existence, and *kāya*, that are capable of occupying space or are related to many spatial points.

The six substances (*dravyas*) are:

1. *jīva* (conscious being),
2. *ajīva*, i.e. non-conscious matter (*pudgala kāya*) or non-souls,
3. medium of motion (*dharma*),
4. medium of rest (*adharma*),

5. space (*ākāśa*), and
6. time (*kāla*).

Of the six *dravyas*, the first five are *astikāya* (*PKS* 4). They are eternal, uncreated and of existence magnitude (*Ratnakaraṇḍa Śrāvakācāra* [*RKS*] 4). Whatever things have the essential nature of manifesting themselves severally through their numerous qualities and modes are of the *astikāyas*. These fill the three worlds and are the constituent parts of the worlds (*RKS* 5).

Of the six substances, the sixth, i.e. time (*kāla*) is not included in *astikāya*. It (time) has neither potential no really the capability of the possibility of such space relations. It is denied the name of *astikāya*, and as such it is not included in *astikāya*. It (time) is purely an entity of mono-dimensional series. Though it is not *astikāya*, it does not cease to be a real entity, existent reality (*dravya*) (*PS* 95).

Another name or title of this treatise is the *Pañcāstikāyasāra,* i.e. a collection (*samavāya*) of five *astikāyas*, which is mentioned in verses 103 and 173. In both verses, the title of *Pañcāstikāyasāra-saṁgraha* is qualified by *Pavayanasāra* (*Pravacanasāra* in Sanskrit) meaning the essence (*sāra*) of the Divine Word, utterances or discourses. Thus, the *Pañcāstikāyasāra* is the popular name of this treatise.

The five *astikāyas* together with time (*kāla*) form the six substances (*dravyas*) (*PKS* 6). The very term "substance" is suggestive of something that which subsists or persists in spite of the changes, i.e. permanence through change. The existent reality (*sat*) is the characteristic of *dravya* (substance) (*TS* V.29). It is characterized by origination, destruction and permanence. It is also the substratum of its qualities (*guṇas*) and modes (*paryāya*) (*PKS* 10).

Though these six substances are mutually interpenetrating and accommodating one another and though they commingle and get mixed up in view of occupying the same space, yet they always maintain their own intrinsic nature (*svaka svabhāva*) (*PKS* 7).

After describing substances in general in 21 verses (*PKS* 6-26), Kundakunda examines in detail *jīva dravya* (soul substance); in 47 verses (*PKS* 27-73), *pudgala dravya* (matter substance); in nine verses (*PKS* 74-82), *dharma-adharma dravya;* in seven verses (*PKS* 83-89); and *ākāśa* (space) in 15 verses (*PKS* 90-104) in four chapters of Book 1.

In *Pañcāstikāyasāra* 105, again bowing his head to Lord Mahāvīra, the cause of liberation, Kundakunda describes nine principles (*tattvas*) or categories (*padārthas*) [the first two primary categories of *jīva* (soul, i.e. conscious being) and *ajīva* (non-conscious, non-living thing, including matter (*pudgala*), and the seven derivative categories of merit or virtue (*puṇya*), sin or *pāpa*, influx (*āśrava*), stoppage (*saṁvara*), shedding (*nirjara*) and bondage (*bandha*) of *karmas* and liberation (*mokṣa*) as also the path of liberation (in two verses – *PKS* 106-107)]. The nine categories are mentioned in *Pañcāstikāyasāra* 108 and discussed in eight chapters (in *PKS* 105-153). Then there is resume (*cūlikā*) of expansion or development of the path of liberation (*PKS* 154-173).

The enlightened conduct united or combined with right knowledge, devoid of or uninfluenced by attachment and aversion, forms the path of liberation to those faithful souls who realize intelligence, or the sense through the (five) attainments (*labdhi*) (*PKS* 106).

(The evil or inauspicious (*aśubha*) thought-activities of) carelessness or passion-infected thoughtless conduct (*pramāda-caryā*), impure emotions, hankering or craving for sensual pleasures, causing anguish to fellow beings and slandering them, openly or covertly, constitutes the inflow (*āśrava*) of evil or sin (*pāpa*) (*PKS* 139). Commendable or excellent attachment (*praśasta rāga*) auspicious (*śubha*) thought-activities compassion or sympathy (*anukampā*) and absence of impurity in the mind – these three are the auxiliary causes (*nimitta kāraṇa*) of the inflow (*āśrava*) of virtue or merit (*puṇya*) in *jīva* (soul) (*PKS* 135, 135 *AC*). The dawn of enlightened world-view (*samyak darśana*) presupposes the acquisition of five *labdhis* (attainments).

Conviction or belief in the existence and reality of (nine categories, *padārthas*) or spiritual principles (*tattvas*) is enlightened world-view (*samyakatva* or *samyak darśana*); knowledge of their real nature [without doubt or error] is enlightened or right knowledge. And the attitude of equanimity ensconced as revolving or use on the path is enlightened or right conduct. These three together constitute the path (of liberation) (*PKS* 107).

Other Works of Kundakunda

Besides the *Pañcāstikāyasāra*, there are four other works of Kundakunda. They are the *Pravacanasāra* (*PS*), *Niyamasāra* (*NS*), *Samayasāra* (*SS*) and *Bāras Anuvehā* (*BA*).

Except *Bārasa Anuvekhā*, all the other four works of Kundakunda are ending with the suffix *sāra* (essence). The suffix word *sāra* is affixed to the first four works of Kundakunda in order to avoid any perverse, i.e. distorted, meaning or the real sense (*vipariya pariharitham* in Prākr̥t and *viparīta pariharaṇam* in Sanskrit (*NS* 3)).

The first verse of all the five works of Kundakunda invariably pays obeisance or homage to Lord (Jina). Thus, *Pañcāstikāyasāra* 1 and *Niyamasāra* 1 pay homage to the Omniscient Lord (Jina). *Samayasāra* 1 and *Bārasa Anuvekhā* 1 pay obeisance to all the liberated souls (*savva siddhas*). And *Pravacanasāra* 1 pays homage to Lord Mahāvīra (Vardhamāna), the twenty-fourth *tīrthaṅkara* (*PS* 1).

In addition to Lord Mahāvīra, *Pravacanasāra* also pays homage to other *tīrthaṅkaras*, who are the promulgators of the order or the path of *dharma*, i.e. righteousness and piety, and all the *siddhas*, and the saints (*śramaṇas*) of the pure nature, who are engaged in the practice of enlightened knowledge, enlightened world-view (*darśana*), enlightened conduct, austerities and vigour or energy (*vīrya*) (*PS* 2). Kundakunda also pays homage to all the *arhats* (*jīvan-mukta*, i.e. liberated souls in embodied existence) as compared to disembodied liberated souls, the *siddhas*) (*PS* 3).

Kundakunda also pays homage to *gaṇadharas* (apostles or direct disciples of the *tīrthaṅkaras*), the group of *adhyāpakas* (*upādhyāyas*, the spiritual teachers of saints) and all the saints (*sādhus*) (*PS* 4). Having approached their holy station, devoted primarily to purifying *darśana* and *jñāna*, he then relies upon equanimity (*sāmya*), leading to emancipation (*nirvāṇa*) (*PS* 5), which is reached by means of a conduct based on (enlighted) *darśana* and *jñāna* (*PS* 6).

In *Pravacanasāra* 7, Kundakunda states: "Conduct is, indeed, *dharma*; *dharma* is defined as equanimity and equanimity is a state-of-*evolution* (*pariṇāma*) of the Self, which is free from infatuation (*moha*) and disturbance (*kṣobha*)", caused by attachment and aversion.

Another distinguishing feature of Kundakunda's works is that they all speak of *mokṣa-mārga* (path of liberation) and liberation (*nirvāṇa*). Thus, *Niyamasāra* 2 and *Pañcāstikāyasāra* 2 speak of *nirvāṇa*; refer to *nirvāṇa* (in *Pravacanasāra* 5 and 6); (*Bārasa Anuvekhā* and *Samayasāra*) speak of *nirvāṇa* and *mokṣa-mārga* (path of liberation) towards the end (in *BA* 91 and *SS* 410-412 and 414). It is interesting to note how the constituents of *mokṣa-mārga* are described in his different works.

Kundakunda promises to narrate *Samaya-Pāhuḍa*, i.e. the *Samayasāra* uttered or spoken (*bhaṇitam*) by all-knowing masters of scriptures (*Śrutakevalī*) (*SS* 1) and to narrate *Niyamasāra* uttered or spoken by the Omniscient Lord (Kevalī) and *Śrutakevalī* (*NS* 1). He speaks of narrating *Pañcāstikāyasāra* on the basis of the scriptures (*samayāmaika*, i.e. *āgama*) revealed by the Jinas (*PKS* 2). He implies to narrate *Pravacanasāra* on the basis of *dharma tīrtha* (law of righteousness and piety) promulgated by the Spiritual Guide Lord Mahāvīra (*PS* 1). And he speaks of narrating *Bārasa Anuvekhā* by relying on excellent concentration or meditation (*BA* 1).

Yet another distinguishing feature of Kundakunda's work *Pravacanasāra* is that it speaks of three kinds of *upayoga: śuddha* (pure), *śubha* (auspicious) and *aśubha* (inauspicious) (which are defined in *Pañcāstikāyasāra* 64, 69 and 157-158) and nine *tattvas* or *padārthas*, including merit (*puṇya*) and sinful (*pāpa*), thereby distinguishing *śubha* or *puṇya* from *aśubha* or *pāpa*. Kāñjīpanthī, however, emphasize only two conscious attentiveness (*upayoga*), viz. *śuddha* (auspicious) and *aśuddha* (inauspicious) [*puṇya* (merit) or *śubha* (auspicious), *puṇya* (*pāpa*) or inauspicious (*aśubha*) are both included in *aśuddha upayoga*]. They also speak of seven *tattvas*, thereby including boh *śubha* and *aśubha* in inflow (*āśrava*) of *karmas*.

Mithyātva (deluded view), *ajñāna* (ignorance or nescience) and *avirati* (perverse conduct) are the three perverted *bhāvas* (of the Self), which are opposite of enlightened world-view, right knowledge and right conduct, which together constitute the path of liberation.

Like *mithyātva*, three kinds of *upayoga* (the term *tinni* or *tiviho* is mentioned in *Samayasāra* 89-90 amd 94-95) are the beginningless states-of-evolution (*pariṇāma*) (of the Self) (*SS* 89). Whatever psychic state or mode (*bhāva*) (the Self) does, *upayoga* becomes the causal agent or doer (*kartā*) of that *bhāva* (*SS* 90), which could be good (*śubha*), evil (*aśubha*) or pure (*śuddha*) (*PS* 9).

PRAVACANASĀRA

The *Pravacanasāra*, the treatise on Divine Discourses or the Essence of Jaina Scriptures, is primarily concerned with the investigation and consideration of knowledge (*jñāna*) and knowables (*jñeya*). With a view to save the knowledge of the blemish (*kalaṅka*) and corrupting influence (*dūṣaṇa*), *jñāna* is described as indeed the nature of self-sensation or intuition and really different or distinct from the states-of-evolution of appearance or form of knowable (*jñeyakasāra*), and of the nature of self-intuition (*sva-saṁvedana*). This investigation and consideration of knowledge and knowable is the subject matter of *Pravacanasāra*.

The Pravacanasāra is divided into three books:

1. The Principle of Knowledge (*Jñāna*) containing an excellent exposition of knowledge;
2. the Principle of the Knowable (*Jñeya*), which describes the

substance in general, and substance in particular, as also distinctive features of *jīva* (soul) substance; and

3. conduct which narrates the path of liberation and the dual role of the good, virtuous or merit (*puṇya*) or auspicious (*śubha*) psychic-attention, as also the Five Jewels in the end.

(The doctrine is that) knowledge is the Self (*ātmā*). There is no knowledge without the Self; therefore knowledge is the Self. But the Self is knowledge and something else [also] (*PS* 27), such as intuition (*darśana*) volition, bliss and vigour. There is, thus, unity or oneness (*ekatva*) as well as otherness (*anyatva*) between the Self and knowledge (introductory remark of Amr̥tacandra to *Pravacanasāra* 27).

Commenting on the verse, Amr̥tacandra states:

> There is no existence of knowledge without the Self – for knowledge arises, requiring and using as basis, the Self alone, which shows towards it a beginningless and endless relation of inherence, ensuing from its innate nature. There is no tendency towards such a relation of inherence with reference to all other non-conscious things (*vastu*). Therefore, we may say that knowledge is the Self.

In the next verse (*PS* 28), it is stated that the knower possesses knowledge as his innate nature; the objects, for sure, possess the character of knowables with reference to the knower, like shapes and forms with reference to the eye. But they do not abide in each other (*PS* 28). Even without mutual penetration, the Self and the objects are ready to receive and to occasion the appearances of all knowables without being identical with one another (*PS* 28 AC).

Giving the definition of substance (in Book II of *Pravacanasāra*), Kundakunda states:

> That which, whilst it does not forsake its innate nature, is characterized by origination, annihilation and stability and which possesses or is endowed with qualities and modifications, is called substance. – *PS* 95, also in *TS* V.29-30 and 38

The Self (*appa* or *ātmā*) is of the nature of *upayoga* (conscious attentiveness, psychic-attention or manifestation or function of consciousness, i.e. the state-of-evolution in obedience to, acting in accordance with or in conformity/compliance with consciousness) (*PS* 155). I (*aham*) is *upayoga* alone (*SS* 36-37); *upayoga* is in *upayoga*, not in anger, etc. (*SS* 181); and not in material *karmas* and no-*karma* (physical body) (*SS* 182). And when the soul happens to possess unperverted cognitive activity (*upayoga)* free from doubt and error, i.e. based on the intrinsic pure nature of consciousness, it does not indulge in any activity or thought contrary or opposite to the innate pure nature of the Self (*SS* 183).

In *Pañcāstikāyasāra, upayoga* is considered a particular (*viśeṣa*) form, property or aspect of consciousness. *Upayoga* of the Self is of two kinds: (i) good, virtuous, meritorious or auspicious (*śubha*) or *puṇya*, and (ii) evil, vicious, sin or inauspicious (*aśubha*) or *pāpa* (*PS* 155-156). In the *Pravacanasāra*, *upayoga* is considered to be of three kinds: pure (*śuddha*), auspicious (*śubha*) and inauspicious (*aśubha*), which are defined in verses 14, 69 (also *PS* 157 and 158 respectively).

In the *Samayasāra*, *upayoga* when united or combined with infatuation (*moha*) [i.e. *mohanīya karma* both view-deluding (*darśana mohanīya*) and conduct-deluding, i.e. attachment and aversion)] is known as of three kinds: *mithyātva* (delusion), *ajñāna* (perverted knowledge) and *avirati* (lack of moral conduct of self-restraint (*saṁyama*)).

However, *upayoga* is not the same as consciousness which always remains one (*eka*), ever-existent, pure, substance in general (*dravya sāmānya*) and permanent or eternal (*śāśvata* or *dhruva*). *Upayoga* is many, i.e. intuition and knowledge, can be good or evil, and pure or impure as also *viśeṣa* (particular or mode) of consciousness. When impure *upayoga* is the cause of the soul's conjunction with other substance (*PS* 155-156 AC), *Samayasāra* 173 also states that with the exercising of *upayoga* the activity of the soul through the psychic dispositions such as attachment, the previously bound *pratyaya* of passions like anger (*SS* 109), become the cause of bondage of *karmas*.

I am not other substance, such as body, mind and speech (*PS* 160).

Inauspicious psychic-attention (*aśubha upayoga*) is extremely worth abandoning (*heya*) (*PS* 12 AC). Moreover, whereas infatuation (*moha*) and aversion are evil or inauspicious (*aśubha*), attachment is both *aśubha* and *śubha* (auspicious) (*PS* 180). *Śubha* has a purifying (*viśuddhi*) effect and *aśubha* is imbued with affliction and anguish (*PS* 155 AC and 180 AC). The conduct of excellent attachment gradually leads to the highest happiness of liberation (*PS* 254 and 254 AC).

In *Pravacanasāra* 192, Kundakunda highlights seven significant distinguishing features of the innate pure nature of the Self in the following words:

> I deem the Self as pure because it is endowed with qualities or the nature of knowledge (*jñātamānam*), and of intuition (*darśana bhūta*), suprasensual (*atīndriya*), the great thing or the supreme reality (*moha-artha*), eternal (*dhruva*), stable (*acala*) and without any prop or support (*anālamba*).

In the last five verses (*PS* 271-275), Kundakunda describes the reality of the five principles (*tattvas*), called the five jewels by Barend Faddegon.[1] These five principles are:

1. *Reality of Transmigration Principle* (*saṁsāra-tattva*): Those who in (comprehending) the scriptures (*samaye*) grasp the reality of things not as they are or wrong or perverted perception of the reality of things, i.e. other than as they really are; being convinced (in their mistaken way) that "so and so are the reality", wander through all later time (in the transmigratory existences), which brings them an endless fullness of kārmic fruit (*PS* 271).

[1] Barend Faddegon (ed.), 1935, *Pravacanasāra*, with an Introduction by F.W. Thomas, Jaina Literature Society Series, vol. 1, Cambridge: Cambridge University Press.

2. *Reality of Liberation Principle* (*mokṣa-tattva*): He who is separated or different/distinct from the improper conduct, who has convictions concerning the nature of the reality of things (*padārthas*) as they are in accordance with reality, real, actual, genuine, true and realistic (*yathārtha*), whose Self is calm or pacified and reposeful and whose sainthood is consummated, does not live long in this unfruitful world here [in this transmigratory existence] (*PS* 272). In other words, he who is endowed with enlightened world-view of the reality of things as they are attains liberation in a short time.
3. *Realilty of the Principle of the Means of Liberation* (*mokṣa-tattva-sādhana-tattva*): Those who have known or perceived the reality of things or categories in an enlightened, right or proper way (*samyak*) and who have given up, abandoned, renounced or discarded external and internal *upādhis*, i.e. the limitations of kārmic adjuncts and encumberances, and who are not attached to sense-objects, are designated pure (*śuddha*) (*PS* 273).
4. *Praising or Extolling the Reality of the Principle of the Means of Liberation:* He, who possesses the innate pure nature of the Self, is spoken of as a saint (*śramaṇa*). He alone is endowed with (right) intuition (*darśana*) and knowledge; he alone attains liberation (*nirvāṇa*) [whose divine innate nature is imbued with knowledge and bliss]. He alone becomes a liberated soul (*siddha*). Homage to him (*PS* 274).
5. *Reality of the Principle of the Fruit of the Treatise:* He who, practising the conduct of a householder or homeless (properly) comprehends or understands this scripture, acquires in a short time, the essence (*sāra*) of the (Divine Discourse(s) or the scripture *Pravacanasāra* 275, i.e. he realizes the ever-existent reality (*bhūtārtha*) (which exists in all living beings in all stages of life's journey), [whether transmigratory

existence of human, subhuman (animal) mode or in the liberated state], the intuitive divine knowledge and blissful nature (*svabhāva*) unprecedented (*abhūtapūrva*) glorious divine supreme reality of the Self (*bhagavanta ātmā*) (*PS* 275 AC).

In the Appendix concerning 47 points of view regarding the Self, Amr̥tacandra states: "What is this Self and how is it obtained or attained?" He observes:

> In the first place, the Self is one substance having endless or infinite properties, pervaded by the generality of "consciousness" because it is trustworthy known by self-experience (*sva-anubhāva*), preceded by the valid knowledge characterized by the scripture knowledge pervaded by the infinitely many points of view (*naya*), which in their turn pervade the infinitely numerous properties (*dharmas*).

1. According to "the point of view of substance" (*dravya naya*), it (the Self) is only consciousness, like a simple piece of cloth.
2. According to "the point of modifications" (*paryāya naya*), it is only intuition, knowledge, etc. i.e. it is of the nature of various forms from *paryāya naya*.
3. From "the point of view of existence or considering (things) as existent" (*astitva naya*), it is existent with respect to its own (a) substance (*dravya*), (b) place (*kṣetra*), (c) time (*kāla*), and nature or character (*bhāva*), i.e. something of the nature of knowing things.
4. According to "the point of view or method of non-existence" (*nāstitva naya*), it is with reference to (a) other substance, (b) other place, (c) other time, and (d) other nature or character (*bhāva*), it is something that possesses non-existence.
5. From "the external, other-referential point of view" (*vyavahāra naya*), the Self is that which undergoes bondage of and liberation from other substance, i.e. matter (*pudgala*) [or material *karma*].

6. From "the internal, self-referential point of view" (*niścaya naya*), the Self itself undergoes bondage and liberates it through its own attachment, etc. psychic dispositions or absence thereof.
7. According to "the point of view of the impure" (*aśuddha naya*), it is something having the nature of connection with *upādhi*, i.e. kārmic adjuncts and encumberances, of the character of attachment and aversion.
8. According to "the point of view of the pure" (*śuddha naya*), the innate pure nature of the Self is something having a nature of *nirupādhi*, i.e. devoid of or free from the limitations of *karma upādhi*.

After thus expounding the soul-substance (from various points of view, leaving aside 39 not so relevant *naya*), Amṛtacandra describes the means of the attainment of the pure nature of the Self in the following words:

1. The psychic state of beginningless infatuation [both view-deluding and conduct-deluding, i.e. attachment and aversion] in which material *karma* acts as an indirect extrinsic supportive factor (*nimitta*) [or auxiliary cause (*nimitta kāraṇa*)];
2. the evolution through modes of cognition that are successive;
3. making friendship with particular external knowable things due to the *nimitta* of cognitions;
4. exclusively turned outwards through weakening of the discriminative insight (*viveka*) of the Self; and
5. following after the duality of attachment and aversion, which are causes of (knowledge-obscuring, etc.) material *karmas* – these five things keep the attainment of the pure nature of the Self at a distance.

But when (1) due to the destruction of infatuation, which is the producer or the cause of material *karmas*; (2) distinguishing itself from the changing particular modes of cognitions and remaining undisturbed; (3) having no more friendship with particular external, knowable things; (4) exclusively directed towards discriminative insight (*bheda-vijñāna*) between of the Self and other; and (5) far from following the duality of attachment and aversion leading to the accumulation of material *karmas* – these five lead to the attainment of the glorious, divine supreme reality of the Self (*bhagavān ātmā*) of the nature of unprecedented or extraordinary knowledge and bliss, never before realized.

Amr̥tacandra concludes his commentary, *The Lamp of Truth* (*Tattvadīpikā*) of the *Pravacanasāra* with the words:

> May your consciousness experience it to the full, since here (in this universe) is only consciousness and nothing else; the one supreme reality is consciousness alone (*PS Kalaśa* 22) [i.e. non-dual (*advaita*), without a second].

When Amr̥tacandra speaks (1) in his description of twenty-fourth of the 49 *nayas* in the Appendix that "According to the point of view of non-duality (*advaita naya*) of knowledge and knowable (*jñāna* and *jñeya*), it (the Self) is one (*eka*), like fire, when evolved through a great mass of fuel"; (2) in (*Pravacanasāra Kalaśa* 22 to the effect that "here (in this universe) there is only consciousness (alone) and nothing else", i.e. it is one without a second, and that there is "one supreme reality" of consciousness (alone); and (3) in *Samayasāra Kalaśa* 9 that when one experiences the state of the Self from which all the differences or distinctions of *nayas* (points of view), etc. are ended or extinguished, at that time the duality (*dvaita*) does not make appearance or does not even appeal (at all), he sounds or appears like Advaita Vedāntin; but his Advaita does not assign unreality to the other forms of existence and knowledge, i.e. the objective reality of the world consisting of individual selves and material objects.

Amr̥tacandra's Advaita is not the same as Advaita Vedānta which speaks of "one and only one Absolute, monistic, unchanging reality of universal or cosmic *Brahmn*. Such a concept of an all-encompassing Self, such as the Vedāntic *Ātman* or *Brahmn*, J. Krishnamurti observes, is "just another thought-construction and another manifestation of illusion".[2]

NIYAMASĀRA

Niyamasāra 2 states:

> The path or the way (*mārga*) and the fruit (i.e. the result, outcome or consequence) of the path, i.e. *mārgaphala*, are the two integral parts. *Mārga* is the means of liberation (*mokṣa*) and its fruit is emanicipation (*nirvāṇa*).

Firm conviction (*śraddhā*) in the Jaina *tīrthaṅkara* (as true God) (*āpta*) nature of the Self, scripture (*āgama*) and the true nature of the basic spiritual principles (*tattvas*) is the enlightened world-view (*samyakatva* or *samyak darśana*). He who is free from all defects, blemishes and faults, and is endowed with all the essential qualities, is the (adorable) *āpta* (*NS* 5).

[Also] conviction (in the spiritual principles or things (*tattvas*) as they are and devoid of indiscriminate clinging to wrong or perverse views is *samyakatva*. And (knowledge) that is devoid of doubt (*saṁśaya*), bewilderment or infatuation (*vimoha*), and confusion or vacillation (*vibhrama*) is enlightened knowledge (*samyak jñāna*) (*NS* 51).

Moreover, conviction, which is devoid of or free from fickle, unsteady, transient or ephemeral (*chala*), impurity (*mala*) and feeble (*agādha*) is alone the enlightened world-view (*samyakatva*); correct knowledge (*adhigama*) of what is worth renouncing (*heya*) and what is worth acquiring (*upādeya*) among the *tattvas*, i.e. the basic spiritual principles or reality of things as they are, is enlightened knowledge (*NS* 52).

Niyamasāra 38 states:

---

[2] See Hillary Rodrigues, 2001, *Krishnamurti's Insights*, p. 73, Varanasi.

> The embodied living beings (*jīvas*), etc. are external principles (*bahi tattvas*); they are *heya*, i.e. worth renouncing and not to be aspired for, because they are the attributes and modes caused by the limited conditions of *karmas* (*karma upādhi*). The only thing worth acquiring (*upādeya*) is one's own Self (*ātmano-ātmā*) (*NS* 38), i.e. one's own innate pure nature of the Self.

The *Niyamasāra* is primarily composed to make people understand the conduct of a person from external, other-referential (*vyavahāra*) and internal self-referential (*niścaya*) perspectives or points of view. One should avoid or eliminate inauspicious (*aśubha*) thought activities (*NS* 66) and practise moral self-restraints of non-violence, etc. by observing *vyavahāra* conduct.

Vyavahāra conduct, consists of five vows (*vratas*) of non-violence, truth, etc. or moral self-restraints, five self-regulations or carefulness (*samiti*) in walking, speaking, eating and so on, three restraints of mind, speech and body (*guptī*) and adoration of five most revered spiritual dignitaries (*pañca parameṣṭhi*), is described in 20 verses (*NS* 56-75).

Niścaya conduct, consisting of repentance of past misdeeds (*pratikāman*), not to commit misdeeds in future (*pratyākhyāna*), confession of misdeeds in the present (*ālocanā*), expiation (*prāyaścitta*), profound meditation on equanimity (*samādhi*), enlightened or right devotion (*bhakti*), the essential, self-disciplinary activities (*anaśyaka*) is discussed in 82 verses (*NS* 77-158). In the end, pure conscious attentiveness is described in 29 verses (*NS* 159-187).

Niyamasāra 102 describes the innate pure nature of the Self in these words:

> I am *aigo* (*eka*, one alone, solitary), permanent (*śāśvata*) Self (*appa*), characterized by intuition (*darśana*) and knowledge (*jñāna*). All other remaining states or modes (*bhāva*) are external *bhāvas*, since they all have the characteristic of joining together (*saṅjoga* or *saṁyoga*) [of two substances].

BĀRASA ANUVEKHĀ (TWELVE ANUPREKṢĀ)

Even though (*Bārasa Anuvekhā*) is a short treatise of 91-92 verses, it brings into focus the reality of the permanent (*śāśvata*, *BA* 7 and *NS* 102) "I" beneath the ever-changing, transitory apparent "I" of the body and the senses (*BA* 6).

The treatise emphasizes that the supreme reality of the Self (*āda* or *ātmā*, the conscious, sentient principle), which is distinct or apart from the limitations of the bondage of *karmas* and their fruition, etc. is the only protection, shelter or refuge from births, old age, death, disease and fear (*BA* 11).

The five revered Supreme Spiritual Dignitaries (*parameṣṭhi*): Arihanta (Omniscient Lord), *siddha* (liberated soul), *ācārya* (head of saints), *upādhyāya* (teacher of saints) and saints (*sādhus*) are the five exalted, most revered and adorable Supreme Dignitaries (*parameṣṭhi*) of Jainism, characterized by 46, 8, 36, 25 and 28 qualities respectively.

The five *parameṣṭhi* are said to be the protection or shelter due to being immersed, engrossed or absorbed in their own pure nature of the Self, hence they are considered protection for me (*BA* 12). Moreover, they are described as protection from the external, other-referenial point of view (*vyavahāra naya*). In reality, they are the particular condition or particular mode (*avasthā viśeṣa* or *paryāya viśeṣa*) of only the Self (*ātmā*) alone. Consequent upon the evolution of that condition, the said supreme, glorious, divine to be adored (*bhāgavata* or *bhagavān*) condition is one's own (*svayam*) illuminated; then the five *parameṣṭhi bhagavān* are said to be the protection.[3]

In view of the above, the enlightened world-view or right belief (*samyakatva*), right knowledge, right conduct and right austerities (*sat tapa*). These four are the features, qualities being the result of the soul's own efforts or attempts that inhere, exist or are comprised in the Self (*āda* in Prākr̥t and *ātman* in Sanskrit)

---

[3] Devendra Kumar Shastri, *Bārasa Anuvekhā*, p. 13.

itself alone; therefore only the Self alone is my protection, shelter or refuge (*śaraṇa*) (*BA* 13) from the internal, self-referential point of view (*niścaya naya*).

Like *Samayasāra* 38 and 73 (*Bārasa Anuvekhā* describes *aham* (I) as *ekko* (*eka*), one, alone, solitary, unmixed) and hence "pure" (*śuddha*) nature, characterized by intuition and knowledge, and that the self-restrained person should contemplate that oneness (*ekatva*) with one's own innate pure nature of the Self is worth acquiring (*BA* 20).

In order to remove the mistaken view of Kañjīpanthī that good, virtuous (*puṇya*) or auspicious (*śubha*) is, like sinful (*pāpa*) or inauspicious (*aśubha*), the cause of only bondage of *karmas*, *Bārasa Anuvekhā* categorically states that good or auspicious activity of mind, speech and body (*yoga*) is instrumental in the stoppage (*saṁvara*) of the inflow of evil or inauspicious activities of mind, etc. (*aśubha yoga*). The prohibition of auspicious thought-activity is possible by means of pure thought activity (*BA* 63).

And since that which is the cause of stoppage of *karmas* is also the cause of shedding (*nirjara*) of *karmas* (*BA* 66), *śubha* is also the cause of shedding of *aśubha karmas*. *Saṁvara* and *nirjara* are the means of liberation.

Of course, pure conscious attentiveness (*śuddha upayoga*) is the most desirable thing and is indeed our refuge, but so long as one does not attain *śuddha upayoga*, one has to rely on the *vyavahāra* conduct or auspicious (*śubha*) activity.

The *Bārasa Anuvekhā* contains reflections on the concept of *ekatva–anyatva*, i.e. unity or oneness (*ekatva*) [with one's own innate pure nature] and otherness (*anyatva*) [with body, wealth, etc.] (*BA* 14-23), or *ekatva–vibhakta*, which is the subject matter of *Samayasāra* 5, laying emphasis on the distinctness (*vibhakta*) between one's own substance (*svaka dravya*) and other substances (*para dravya*) (*BA* 85). The *Bārasa Anuvekhā* also speaks of the concept of *heya–upādeya* (*BA* 86) (what is worth abandoning or renouncing (*heya*) and what is worth acquiring (*upādeya*) by stating that the impure psychic states or modes (*paryāya*) caused by the coming into operation or rise (*udaya*) of (material) *karmas* are *heya* and that enlightened knowledge (*sat* or *samyak jñāna*), which is one's own (*svaka*) (nature of) the Self substance (*BA* 84). (This verse is similar to *Samayasāra* 38.) Thus, the *Bārasa Anuvekhā* emphasizes the importance, significance and rarity of discriminative insight (enlightenment) or *bheda-vijñāna*.

In the end, the *Bārasa Anuvekhā* declares that all the living beings, who have attained liberation since beginningless time have attained that status due to the efficacy of twelve contemplations alone (*BA* 89).

The significant points and distinguishing features of the four works of Kundakunda (viz. the *Pañcāstikāyasāra*, *Niyamasāra*, *Pravacanasāra* and *Bārasa Anuvekhā*) have been described and discussed above in brief as they are most relevant and useful in understanding and comparing the significant and important concepts in the *Samayasāra* treatise.

DISTINGUISHING FEATURES OF SAMAYASĀRA

The *Samayasāra*, "the most outstanding treatise in the field of spirituality", has a number of distinguishing features. In brief, they are:

1. Spirituality is concerned with the spirit (the immaterial being), soul, living soul, the Self, sentient being and consciousness. The Self (*āda, appa, ātmā*) is said to be conscious, sentient energy, of the nature of consciousness and indeed only consciousness alone (*cinmātra eva*). In spirituality, the focus of life shifts more towards getting personal happiness through mental peace and less on material needs and desires.

2. The *Samayasāra* is replete with the descriptions of what is the Self or soul, and what it is not (*karma*, no-*karma*, attachment, etc.), of which there is a long list (as many as 29 *bhāvas* in verses 50-55, with subdivisions 296 states or modes). They all

belong to the Self from the external, other-referential point of view and are, indeed, used to distinguish the soul from the non-soul, which are distinct from the innate pure nature of the Self.

3. Kundakunda has not only described what the Self is not, but what it is. He has, thus, stated that "I" (*aham*) is *ekko* (*eka*, one, alone, solitary) (*SS* 36-38, 73 and 198-199), and indeed pure (*śuddha*) (*SS* 38 and 73) characterized by intuition and knowledge (*SS* 38 and 73) always incorporeal (*arūpī*) (*SS* 38), devoid of the sense of mineness (*nirmama*) (*SS* 73) and that not even an iota of anything else is mine (*SS* 38).
4. The Self is also declared to be a conscious being (*cheda*) (SS 297), seer or observer (*draṣṭā*) (*SS* 298) and knower (*jñātā*) (*SS* 299), and that all the remaining states or modes are other (*para*) than me (*SS* 297-299).
5. The innate pure nature of the Self is also described as "consciousness of the knowing nature" (*jñeyaka bhāva*) (*SS* 6); *jñeyaka* is verily *jñeyaka* only (*SS* 356); *jñeyaka śuddha*, i.e. *śuddha* due to being *jñeyaka* alone (*SS* 7); *jñeyaka* (*SS* 167, 210-13, 216 and 315); and *jñeyaka svabhāva* (*SS* 200).
6. The role and nature of discriminative insight (*bheda-vijñāna*) (called particular difference or distinction (*viśeṣa antara* by Kundakunda in *Samayasāra* 69 and 71) between one's own (*sva*) and other (*para*) or the soul and the non-soul as also between the Self (*āda*) and anger, etc. influx (*āśrava*) and consequent bondage of material *karmas* (*SS* 69 and 71) is very important because all living beings have attained the state of liberated soul due to *bheda-vijñāna* (*SSK* 131). While Kundakunda has used the term *viśeṣa antara*, both Amṛtacandra and Jayasena have used *bheda-vijñāna* or *bheda-jñāna*. The correct or simple meaning of *bheda-jñāna* is discriminative knowledge between one's own (*sva*) and other (*para* or *anya*).

7. *Bheda-vijñāna* is generally described in three situations: (1) in the determination of *tattva*, i.e. the real state or nature or essence of things, particularly the Self; (2) in the form of capacity or power (*śakti-rūpa*); and (3) it is used as the nature of experience of consciousness, distinct from all other substances. The result or effect of *bheda-vijñāna* is that when *jñāna* (knowledge) is firmly fixed in *jñāna* (i.e. in pure conscious attentiveness, *śuddha-upayoga*) delusion does not recur again; and when *jñāna* is firmly fixed in *śuddha-upayoga*, it is not transformed again in other altered forms.

8. The cognition or understanding of two distinct existent realities of (immaterial) soul and matter (*pudgala*) or material *karma* and no-*karma* (physical body) is mentioned in *Samayasāra* 182. Due to the realm (of thought) of the two being distinct, the two existent realities are not the same.

   When the Self obtains the discriminative and unperverted knowledge (free from doubt, error or contrary and perverted knowledge), the innate pure nature of the Self manifests itself and then the soul ceases to indulge in any kind of impure thoughts (*SS* 183).

9. According to Kundakunda, he, who does not know the innate pure nature of the Self, is an ignorant and unenlightened soul (*ajñānī*).

10. Not knowing the true nature of the Self, the ignorant, unenlightened person, covered with the darkness of perverted knowledge, thinks that the nature of the Self is attachment alone (*SS* 185). And attachment binds the *karmas* (*SS* 150 and *PS* 179).

11. Neither the Self can exist without consciousness, nor consciousness can exist without the Self. Thus, there is relationship of sameness of nature or character between the Self and consciousness (*tādātmya sambandha*). Amṛtacandra says: "Consciousness is verily consciousness alone" (*SSK*

70-89). The real nature of the Self is "all-purity only consciousness alone (*sarva-viśuddha cinmātra*) (*SS* 297 AC); pure of the nature of only consciousness (*SSK* 182); "pure consciousness-as-such" and "one consciousness only" (*eka cinmātra*) (*SS* 298-99 JS).

12. The innate, pure nature of the Self is an ever-existent reality (*bhūtārtha*) (*SS* 11) and pure since *bhūtārtha* is said to be the pure point of view (*śuddha naya*) (*SS* 11) in all three times (*trikālī śuddha*), and without beginning and without end (*SSK* 10 and *PS* 192 AC). The real nature of the Self has unity or oneness (*ekatva*) with its own properties and is only consciousness alone (*PS* 192 AC).

13. The essential nature of the soul substance in general (*sāmānya jīva*) is pure in the sense of non-relational to or independent of *karmas* (*karma nirapekṣa*), not pure in the sense of *karma-rahita* (devoid of *karmas*) as the liberated states of *Arihanta* and *siddha*. The liberated souls (*Arihants* or *siddhas*) are the modes (*paryāya*) of the soul substance. It is completely devoid of or free from *karmas* (*karma rahita*) and hence pure, has a beginning though without an end. The essential pure nature of the Self is the causal (*kāraṇa*) *paramātmā*, while the liberated souls are the effect (*kārya*) *paramātmā*, the glorious divine Lords.

14. Though the living being is associated with *karmas* since beginningless time, his real nature is considered *nirupādhi* pure, i.e. devoid of the limiting conditions of the kārmic adjuncts and encumberances as it is of the nature of non-relational to or independent of *karmas* (*karma-nirapekṣa*) only consciousness-as-such alone (pure *pāriṇāmika bhāva*) (*TS* II.1).

    The other four *bhāvas* (psychic states) are also the Self's own *tattva* (*sva-tattva*), viz. subsidential (*aupaśamika*), destructive (*kṣāyika*) of *karmas*; partly destructive and partly subsidential (*kṣayopaśamika*); and the coming into operation or rise (*udaya*) of *karmas* (*audāyika*) they are all *karma*-referential (*TS* II.1).

15. The real nature of the Self is transcendental to the bias of relative points of view of the Self as being bound (*baddha*) or non-bound (*abaddha*) by *karmas* and in this sense it is transcendental, pure and uncontaminated by *karmas* (*SS* 142 and 144) and that is the essence (*sāra*) of *samaya* (Self), i.e. *samayasāra* (*SS* 142 and 144).

16. Attachment, aversion and infatuation (*moha*) are the *āśrava bhāvas*, i.e. the causes of the inflow (*āśrava*) of *karmas* into the soul, in the absence of which the bondage (of *karmas*) does not take place (*SS* 177).

17. The causal conditions (*hetu*) of attachment, etc. [attachment binds the soul, *Samayasāra* 150 and *Pravacanasāra* 179] are said to be (the manifest operation of) (1) deluded view or delusion (*mithyātva*); (2) ignorance, perverse knowledge or nescience (*ajñāna*); (3) vowlessness, lack of moral self-restraints or non-abstinence from violence, etc. (*avirati*); and the activities of mind, speech and body (*yoga*) – these are called *adhyāvasāna bhāvas* (*SS* 190).

    Adhyāvasāna bhāvas are psychic dispositions that are transposed or superimposed (*adhyāropa*) on what it is not. In other words, it (*adhyāvasāna bhāvas*) is transposition of *mithyātva*, etc. impure *bhāvas* on the innate, pure nature of the Self.

    Śaṅkara defines (*adhyāvasāna* or) *adhyāsa* as the intellect grasping of "that" in what is not "that". This *adhyāsa* corresponds to *viparyāya* (error) of Patañjali (*Yogasūtra* I.8), which means false knowledge which does not correspond to the nature of that (= object), e.g. a rope is seen and understood as a snake.

18. The innate nature of the Self is an ever-existent reality (*bhūtārtha*) (*SS* 11 and 13).

19. *Bhūtārtha* is said to be *śuddha naya*, i.e. pure point of view. *Śuddha naya* is defined in *Samayasāra* 14 in the following words:

> That which perceives Self (*appanam*) as (1) not bound and untouched (*abaddha-aspr̥ṣṭa*) [in the sense of being uncontaminated by the limiting conditions (*upādhi*) of kārmic adjuncts and encumberances; (2) *ananya*, i.e. only one, unique and inalienable from its nature of only consciousness alone in the sense of being distinct from human, hellish, etc. modes (*paryāya*) throughout its transmigratory existences; (3) *niyata*, i.e. constant, steady and invariable in its nature of consciousness [in the sense of remaining same in all conditions of life, and in all stages of life journey in spite of all the changes in the psychic states or in the sensory and scriptural knowledges, etc.; (4) *aviśeṣa*, i.e. uniform, non-difference or indivisible nature of consciousness (in the sense of being transcendental to or no differentiation of knowledge, volition or conduct and *darśana*, though not negating these attributes in the Self; and (5) *asaṁyukta*, i.e. not united or combined [in the sense of being distinct from all other substances, including the non-Self, material *karma*, no-*karma* (physical body) and psychic *karma* (*bhāva karma*), i.e. the psychic states of infatuation (deluded view), attachment and aversion, and anger, etc. passions, in which material *karma* acts as the supportive factor (*nimitta*) or indirect, extrinsic or auxiliary cause (*nimitta kāraṇa*) know that to be *śuddha naya* (*SS* 14).

20. It is said that the inauspicious (*aśubha*) *karma* is bad (*kuśīla*) and auspicious (*śubha*) *karma* is good (*suśīla*), but that which is the cause of the transmigratory existences of the cycle

of births and deaths (*saṁsāra*) how could that be said to be good? (*SS* 145).

Both *śubha* or *puṇya* (virtuous or meritorious) and *aśubha* or sin (*pāpa*) *karma* bind the soul just as the fetters of gold and iron (*SS* 146) and lead to the destruction of the independence of the soul (*SS* 147).

Pravacanasāra 77 is similar to *Samayasāra* 145-147. It states:

> He who does not believe that this alone (*eva*) is that there is nothing particular or special (*viśeṣa*) between virtue or merit (*puṇya*) and sin (*pāpa*) [i.e. there is no particular difference between the two or that the two are similar] moves about, covered with infatuation (*moha sañcānva*) in the terrible cycle of transmigratory existences (*saṁsāra*) of the cycle of births and deaths (*PS* 77).

The word *evam* in this verse signifies that it is to be understood in the context of previous verses. The previous verse (*PS* 74) makes it quite clear that it is only when the prosperity and higher status, etc. acquired as a result of *puṇya karma* makes one "thirst or crave for sense objects (*PS* 74 and *PS* 74 AC) and one takes pride in his superior position or condition, then only *puṇya* is the cause of misery.

And *Pravacanasāra* 76 states:

> The happiness, obtained from the senses, though originating from *puṇya* is indeed or actually misery, as it (1) is dependent on the other (as it rests on something else); (2) is accompanied by obstacles or distractions; (3) is interrupted or terminable; (4) is the cause of bondage; and (5) is unequal for it is utterly unsteady (*PS* 76 and *PS* 76 AC). In these circumstances, *puṇya* turns out to be similar to *pāpa*.

Puṇya is a valuable thing; it should not be squandered in sense objects.

Pravacanasāra 77 does not suggest that "there is no difference whatsovever between *puṇya* and *pāpa*; it only says there is no particular difference". Second, the qualifying phrase *moha-sañchāno* (covered with infatuation is significant because it is primarily due to infatuation) that the soul wanders aimlessly in *saṁsāra*.

Kundakunda and Amr̥tacandra were both aware that there is a considerable difference between *puṇya* and *pāpa* or *śubha* and *aśubha*.

Kundakunda himself says: "*Śubha* gradually brings about the highest happiness of liberation" (*PS* 254). *Śubha*, Amr̥tacandra observes, is endowed with excellent or commendable attachment that has purifying effect (*viśuddhi*) (*PS* 155 AC and 180 AC).

Amr̥tacandra points out that *śubha* differs from *aśubha* on four grounds, viz. (1) in its nature; (2) cause; (3) effect or experience of fruition of *karmas*; and (4) in substratum, being supportive of the path of liberation. *Śubha* is the path of liberation, while *aśubha* is the path of bondage (*SS* 145 AC). Inauspicious (*aśubha*) or sin (*pāpa*) is to be absolutely and completely abandoned as it is worth discarding (*heya*) (*PS* 12 AC).

According to the *Niyamasāra*, one should (1) avoid all evil and abstain from *aśubha* (*NS* 66); (2) engage in moral, i.e. *vyavahāra*, conduct of righteousness (*śubha*) (*NS* 55); and (3) endeavour to purify one's psychic dispositions through self-disciplinary methods of repentance, etc. (*NS* 77-158).

Of the three *āśrava bhāvas* (*SS* 177), aversion and infatuation are *aśubha*, while attachment is both good and evil, because attachment has a double character according as it includes sanity and purification (*viśuddhi*) or morbidity (*saṅkleśa*) (*PS* 180 AC and 155 AC).

Those who are devoid of evil or inauspicious psychic-attention and have pure psychic attention or good psychic attention are saviours of men; he, who is devoted to them, receives excellent merit (*PS* 260).

21. Ācārya Vīrasena in his commentary on one of the most sacred Jaina texts of the Digambara sect *Jai Dhavala* (Book 1, p. 5), viz. the *Kaṣāya-Pāhuḍa* or *Kaṣāya-Prābhṛta*, states:

    If stoppage (*saṁvara*) of the influx (*āśrava*) (and consequent bondage of *karmas*) as also shedding or partial dissociation of *karmas* (*nirjara*) is not considered possible by *śubha* and *śuddha* (pure) psychic states-of-evolution of the Self, then the partial dissociation and complete eradication or destruction of *karmas* will not be possible at all.

22. Kañjīpanthī believe that there are seven basic principles of life (*tattvas*) (*TS* 2.4) as a result of which both *puṇya* or *śubha* and *pāpa* are included in inflow (*āśrava tattva*) of *karmas* and consequent bondage. But Kundakunda categorically mentions nine *tattvas*, including *puṇya* and *pāpa* (*SS* 13 and *PKS* 108), and the *Pravacanasāra* also speaks of (nine) categories (*PS* 264).

23. Kañjīpanthī maintain that there are only two kinds of psychic attention (*upayoga*), viz. (1) *aśuddha* (impure), which includes both *śubha* and *aśubha*, and *śuddha* (pure). But Kundakunda describes (*upayoga*) to be of three kinds (not two) good, evil and pure (*PS* 9). Moreover, in *Pravacanasāra* 155, Kundakunda speaks of *upayoga* as *śubha* and *aśubha*, i.e. evil (*PS* 155). Due to the arising of evil (*aśubha*) the Self becomes a rogue, an animal or an inhabitant of hell and always beset by thousands of pains and strays incessantly (*PS* 12).

24. The Self which has unity or oneness (*ekatva*) with its own innate, pure nature is the most splendid and beautiful ideal in the whole universe. The *ekatva* with bondage (*bandha*) (of *karmas*) is considered to be discordant and problematic, as it accounts for disharmony with that ideal (*SS* 3).

25. The narration of sensuous pleasures and enjoyments (*kāma* and *bhoga*) and bondage (*bandha*) (of *karmas*) have been heard, known and experienced by all. The oneness (*aiyatta* in Prākṛt and *ekatva* in Sanskrit) of the Self with *karmas* is often reproached or scolded (*uvalambha* in Prākṛt), only the distinctness (*vibhakta*), (i.e. the discriminative insight) with them is not easy (*sulabha*) (*SS* 4).

26. The *Samayasāra* is replete with the concept of unity or oneness (*ekatva*) and distinctness (*vibhakta*), which Kundakunda has promised to expound in that treatise (*SS* 5).

27. The relationship of the cause and effect (*kārta-karma*) does not exist between the conscious Self and non-conscious matter (*pudgala*), i.e. material *karma*. The two, however, act as the indirect, extrinsic or the auxiliary cause (*nimitta kāraṇa*) of one another (*SS* 81). Thus, there is a *nimitta-naimittika* relationship between the two, not a *kartā-karma* relationship.

    The impure psychic states of the Self act as the *nimitta* of the transformation of matter (*pudgala*) into material *karmas*, and the material *karma*, in turn, acts as the *nimitta* of the evolution of the Self into impure states (*SS* 80) of attachment, aversion, deluded view or delusion and passion, *pratyaya* or the *adhyāvasāna bhāvas*. Thus, the one acts as the *nimitta kartā* (causal agent or doer) of the other, or one can say that the two act as the cause and effect in the capacity of *nimitta* of each other (*anyonya nimitta*, *SS* 81).

28. The innate nature of the Self is pure (*śuddha*) and untainted or undefiled *bhāva* (psychic disposition) (*SS* 90).

The significant points and distinguishing features of all the five works of Kundakunda (viz. the *Pañcāstikāyasāra*, *Niyamasāra*, *Pravacanasāra*, *Bārasa Anuvekhā* and *Samayasāra*) have been described and discussed above in brief as they are most relevant and useful in understanding the important concepts of Kundakunda.

Select Bibliography

Chakravarti, A., 1971, *Samayasāra of Kundakunda*, Delhi: Bhartiya Jnanapith.

Chakravartinayanar, A., 1975, *Pañcātiskāyasāra: The Building of the Cosmos,* New Delhi: Bhartiya Jnanpith.

Faddegon, Barend (ed.), 1935, *Pravacanasāra*, with an Introduction by F.W. Thomas, Jaina Literature Society Series, Cambridge: Cambridge University Press.

Gier, N.F., 1995, "Ahimsa, the Self and Postmodernism: Jain Vedantist and Buddhist Perspectives", *International Philosophical Quarterly*, 35(1), Issue no. 137.

———, 1996, "Jaina Superhumanism and Gnostic Titanism", *The Scottish Journal of Religous Studies* (Spring and Autumn) 17(1-2).

Jain, Jagdish Prasad, 2005, *Fundamentals of Jainism*, New Delhi: Radiant Publishers.

Jain, Jagdish Prasad (ed.), 2005, *The Enlightened Vision of the Self: Svarupa Sambodhana of Akalanka*, tr. D.K. Goyal, New Delhi: Radiant Publishers.

Jaini, J.L., 1916, *Outlines of Jainism*, Cambridge: Cambrige University Press.

Jaini, J.L., 1974, *Tattvārthasūtra*, New York: AMS Press.

Libet, Benjamin, 2003, "Can Conscious Experiences affect Brain Acitivity?", *Journal of Consicousness Studies*, 10.

———, 2004, *Mind Time: The Temporal Factor in Consicousness,* Cambridge Mass.: Harvard University Press.

Mitra, Kalipada, 1939, "The Jaina Theory of Existence and Reality", *Indian Culture,* January, Calcutta.

Ornish, D., 2001, *Eat More, Weight Less,* New York: Harper Collins Quill.

Shastri, Devendra Kumar,1991, *Barasa Anuvekha* (Hindi), Jaipur: Pandit Todarmal Smarak Trust.

Varni, Jinendra, *Jainendra Siddhanta Kosa*, vol. 4. 2012. Delhi: Bharatiya Jnanapith.

Index